2022

THE FOURTH INDUSTRIAL REVOLUTION

What Every College and High School Student Needs to
Know about the Future

STEPHEN HAAG

Janus Press, LLC

Janus Press, LLC (www.januspressllc.com)

Printed in the United States of America
First Printing: March 2022

Library of Congress
ISBN: 979-8-9857439-0-6

Cover & interior design by riverdesignbooks.com

Janus Press books are available at a special discount for bulk
purchases. For more details, email info@januspressllc.com
or contact the author directly.

For the great leaders in my life.

They include Peggy Aull, Bill Black, and Malnor Arthur (Borger High School), Roy Martin (West Texas State University), L.L. Schkade (University of Texas at Arlington), and James Griesemer, Brent Chrite, and Dan Connolly (University of Denver).

Great leaders enable ordinary people to do extraordinary things.

Table of Contents

THE SETUP

I Want My Pudding Now!

On Demand Is the New *When*

Disclaimers

- I did not harm any animals while writing this book (nor at any other time in my life, for that matter).
- I did not use any GMOs to write this book.
- I am not an attorney-spokesperson.
- Individual results may vary.
- Use only as directed.
- Slippery when wet.
- Batteries not included.
- Void where prohibited.
- Attendant does not have combination to safe.
- Avoid improper microwave use.
- Beware of dog.
- Individual weight loss may vary.
- Contents may be hot.
- Discontinue reading if rash develops.
- Dry clean only.
- Driver does not carry cash.
- Edited for television.
- Employees must wash hands before returning to work.
- Freshest if eaten before date on carton.
- Keep away from fire or flames.

Do not...
- Read this book while operating a motor vehicle, watercraft, or aircraft.
- Remove tag under penalty of federal law.
- Read this book while sleeping.
- Use this book as a personal floatation device.
- Consume alcoholic beverages while operating this book.
- Eat this book. Intended for external use only.
- Read in the shower.
- Leave this book unattended.

Action figures sold separately.

(Whew, glad I got those off my chest.)

Two more…

The subtitle of this book is ***What Every College and High School Student Needs to Know about the Future***. But, it doesn't contain <u>everything</u> you need to know about the future. If that were my talent, I wouldn't be writing books; I'd be playing the stock market and betting sports games. Of course, "betting" wouldn't be the right term, as I would know the outcome of each game before placing my bet.

And finally, this is a book about some of the coolest and hottest technologies out there… artificial intelligence, autonomous vehicles, virtual reality, cryptocurrency, the Internet of Things, quantum computing, and much more. I'm not going to treat them like a black box; instead we're going to delve into some of how and why the technologies work. If you understand the how's and why's, then you can more successfully innovate their use and invent your future.

But, I will be sacrificing some technical accuracy in my explanations to make the technologies more approachable in the small space we have. If you find a place where I have sacrificed technical accuracy for the sake of readability, you have my admiration (for being smart enough to find a technical inaccuracy) and my humblest apology (not real sure why, but it sounded good when I wrote it).

▲ ▲ ▲

Life as (Un)Usual

It's time, everyone. Time to rethink everything we're doing. Time to question the status quo.

We don't need to throw out all the **traditional** way of thinking. Much of it is still good, even exemplary. (To be perfectly honest, some of it was never good to begin with, but let's not open that can of worms right now.)

But as we launch into this time of unbelievable innovation in the 4th industrial revolution, we must reevaluate our approach, our learning, and our thinking, and in some instances perhaps even forego the old for the new.

Of course, the great thing for you is that you're not deeply ingrained with the traditional way of thinking in business, product innovation, and **how things should be because that's the way we've always done them**. But if you are, Alvin Toffler said it best, "The illiterate of the 21st century will not be those who cannot read and write, but those cannot learn, unlearn, and relearn." (Alvin wrote the book **Future Shock** in 1970. Fascinating read. [1])

As you read this particular book, critically think about how you need to approach your education and career choices to best prepare for the future. You still need the basics, but the basics are just the ante to get in the game. To differentiate yourself in the marketplace, you need more. So, as you ponder your future and read this book, grab a sheet of paper and make 2 columns, with the left being "The Basics" and the right being "The More," just as I did on the opposite page. You can see my list, but you're not me. Make your own list. The left column is necessary but doesn't distinguish you from everyone else. The right column is your "bragging rights," how you will market yourself to potential employers, and ultimately how you will define your life.

The Basics...

Personal productivity tools like MS Office

VBA

Zero

Make more money

Why

Create brand

Innovate

Action

Fail fast, learn fast

Personal risk

Organizational reward

2D

Manufacturing

Subtractive

Reduce-Reuse-Recycle (macro)

Transactional data

Sell to the **Fortune 500** and fortunate 500,000,000

The More...

Vuforia

Python

One

Create more value

Why not

Create brand experiences

Disrupt

Creaction

Learn fast, succeed faster

Organizational risk

Personal reward

3D

Harvesting

Additive

reduce-reuse-recycle (micro)

Sensed data

Sell to the unfortunate billions

Some of these terms may be unfamiliar to you right now, and that's okay. Just keep them in mind. I'll point you back to this page from time to time.

▲ ▲ ▲

Pudding on Demand

The ol' fashioned way of making pudding involved heating and whisking milk and egg yolks, constantly stirring, slowly adding cornstarch, sugar, and salt, adding vanilla as the last ingredient, and then refrigerating it for several hours. It was by no means *instant*.

Of course, now we have instant pudding.

And we live in an *instant pudding society*. Everyone seems to want everything right now. The likes of Amazon, Walmart, and all major grocers and big box retailers and delivery services have facilitated that thinking. Order from Amazon today and get it tomorrow or perhaps even today.

Scroll through movies using your cable provider, Amazon Prime, Netflix, etc., select the movie you want and watch it. Download an e-book and read it right then. Order up your favorite fast food and DoorDash or Uber Eats will deliver it shortly. The list goes on and on.

The 4th industrial revolution will bring about the ability to do and have even more on demand.

- **Books on demand** – We do already have this with e-books but the traditional bricks-and-mortar bookstores are hopping on the bandwagon. Select the book you want from the shelf, scan the QR code, and put the book back on the shelf. Proceed to the counter where you'll find your book, printed on demand in the back room. (No inventory, just one of everything.)
- **Food on demand** – Delicious, flavorful, and healthy 3D-printed foods are becoming popular with 3D printers. I suspect you'll add cartridges (just like your printer cartridges for different colors) to your 3D printer and then select from mild, medium, or hot buffalo wings. You'll be able to order these taste cartridges from Chick-Fil-A, Taco Bell, White Castle, and many other restaurants. (BTW, you'll be able to print the food faster than it can be delivered and at a cheaper price.)
- **BBQ sauce on demand** – If you do choose to cook at home and run out of BBQ sauce while grilling, you're just a few voice commands away from having a drone deliver your BBQ sauce to your back deck. (That drone will even find you if you happen to be

grilling at the park or tailgating in advance of the big game. Can you say GPS?)

- **Travel on demand** – Why wait to go somewhere? Just slip into your VR gear and experience any place in the world. Metaverses will become popular vacation destinations... just stay at home and vacation (called staycationing). Some metaverses will start to appear on TripAdvisor. Hearing, seeing, feeling, and *smelling* technologies will be an important part of this virtually real experience. (*Virtually real*... bit of an oxymoron, don't you think?)
- **Transportation on demand** – Owning a car will be replaced by sharing autonomous vehicles. You may someday choose an apartment complex to live in based on its fleet of sharable autonomous vehicles, how many "free" minutes you get per month, response times, etc.
- **Money on demand** – Have instant access to your money with blockchain and cryptocurrency. No more waiting 2 to 5 days for a check to clear. Bank fees will evaporate (yeah).
- **(Back to) Food on demand** – Your refrigerator will sense when items are running low or spoiling and will automatically order them from the grocery store. Your refrigerator will instruct your autonomous vehicle to make the grocery store run (without you).
- **Energy on demand** – Utilizing energy harvesting, you'll have access to electricity no matter where you are, or how far away you are from an electrical outlet. The energy will be free.
- **Sports on demand (of sorts)** – If you don't feel like putting on the uniform and finding a field, play e-sports. It's already a big craze, and 4th industrial revolution technologies like extended reality will truly bring the games to life.

"When" is an important aspect of managing the customer experience, that is, keeping customers happy.

On demand is the new *when*.

So, I'm sure you have many questions. When will all of this happen? How? Why? Can I...?

My short answers: Sooner than you think. I'll explain. Don't ask why, instead ask why not. Yes, you can definitely can.

I know those answers don't satisfy your curiosity. Read on.

Birds of a Feather... You and Me

I envy you, plain and simple. What a great time to be thinking about your future, picking out a career, and chasing your dreams.

Believe it or not, you and I have a lot in common, in many ways. I'm smart and so are you... that's why you're reading this book, and it's also why you're wondering what in the world a young person like yourself has in common with a 60-year-old long-hair professor like me. Thanks for wondering. I'll explain.

When I was about your age, the 3rd industrial revolution was really kicking in gear. That was the late 1970s. We already had large mainframe computers that big organizations were using. I took my first programming class, FORTRAN, on punched cards, in the late 70s. (I knew then that I wanted to be in tech. I was lucky; I was in my teens and I already knew my career path.)

In those late 1970s, we had already moved past the major stages of invention of the 3rd industrial revolution and we were innovating to disrupt. About that same time, the microcomputer explosion began to occur. I bought my first in 1982, a Commodore VIC-20 complete with a cassette tape drive for storage. (No monitor, I had to use my TV.) The birth of the Internet occurred in the 1980s, e-commerce exploded in the 1990s, and the whole SoLoMo thing (social-local-mobile) dominated the first 2 decades of the 21st century.

I was indeed very fortunate. I caught the innovation wave of the 3rd industrial revolution. I was there for the beginning of it and benefited greatly.

Now, here we are at the beginning of the 4th industrial revolution. More appropriately, here you are, ready to catch the innovation wave just I did in the 1970s. And you truly are at the beginning. The tech has been invented, and it's getting better, cheaper, and faster every day.

What's really great for you is that the innovations and applications of the 4th industrial revolution are untapped, unexplored, and ripe for the picking. We haven't really seen anything yet in regards to autonomous vehicles, artificial intelligence, the Internet of All Things, virtual reality, cryptocurrency, metaverses, and many other 4th industrial revolution technologies. True, we have seen the tech, we know it works. Your opportunity is to determine the innovative ways in which the tech will be used.

You and your generation will redefine the competitive landscape of the business world, just as my generation did with innovations like e-commerce.

You and your generation will disrupt ~~almost~~ every industry.

You and your generation have the opportunity to use the technologies of the 4th industrial revolution and permanently close the great divides... economic, quality of life, education, access to health care, and so on.

You will determine the future because <u>you</u> (and take that personally and with pride) will invent it.

So, we are a lot alike. I caught the innovation wave of the 3rd industrial revolution. You get to ride the innovation wave of the 4th industrial revolution.

▲ ▲ ▲

History: Repeat the Good

You've probably heard this saying in some form: Those who don't know history are destined to repeat it. It has kind of a negative connotation, that is, if you know the past, you can avoid making the same mistakes again.

My opposing view is that if you do know history, you can repeat the good parts. I guess I'm a glass-half-full kind of guy.

(FWIW, I like to rethink sayings and proverbs. I think, for example, that you should reinvent the wheel. I also believe that the donut hole is a good thing, because it represents significant opportunity.)

Before we launch into the future, let's take a look at some interesting quotes from the past about the future, what I like to call *notable quotables*.

"Everything that can be invented has been invented."
> Charles Duell, 1899, U.S. Patent Office Commissioner (Since 1900, The U.S. Patent Office has issued more than 11 million patents. Sorry Chuck.)

"Man won't fly for a million years – To build a flying machine would require the combined and continuous efforts of mathematicians and mechanics for 1 to 10 million years."
> *The New York Times*, December 8, 1903 (I really love that one. 9 days later, Orville and Wilbur made history with the first successful flight at Kitty Hawk.)

"I think there is a world market for maybe 5 computers."
> Thomas Watson, 1945, CEO of IBM

"There is no reason anyone would want a computer in their home."
> Ken Olson, 1977, President of Digital Equipment Corp.

"I predict the Internet will soon go spectacularly supernova and in 1996 catastrophically collapse."
  Robert Metcalfe, 1995, Co-Founder of 3Com

"There's no chance that the iPhone is going to get any significant market share. No chance."
  Steve Ballmer, 2007, CEO of Microsoft

Interesting, huh?

Here is my challenge to you. Don't become one of the notable quotables about the future. It's easy – and usually wrong – to make predictions about what won't happen.

Let's, instead, envision possibilities.

To better predict and invent the future, let's take a quick look at the past. After all, we can learn from the past... by not making the same mistakes but more importantly by repeating the successes.

Time for a history lesson. I know, I know... history sucks. But bear with me; it's a fast read. And you get to see 275 years of history in one image.

Agrarian Age

Prior to "industrialization," we lived in what you might call the *agrarian age*, for lack of a better term. It was a time when land was power; generally, the more land you owned the more powerful you were, and not too many people were actual land owners. We use the term "industry" to refer to cottage industries in the agrarian age, people offering  a skilled trade from their homes while still working the land on which they lived. A blacksmith, for example, forged metal products like shoes for horses from their home to provide supplemental income but had to rely greatly on also working the land around the home for food.

Machinery, if you want to call it that, was mainly powered by humans and animals.

Inventory as we know it today was almost non-existent. Instead, we had bespoke products. If you wanted a pair of new shoes, you didn't go into the shoemaker's store and try on different styles, colors, and sizes. The shoemaker took your measurements and made custom shoes to fit you; originally the term ***bespoke*** meant something made specifically for you.

It truly was a time when people lived on the land and off the land. (Small play on words.)

Average life expectancy in the U.S.:
- Men – early to mid 30s
- Women – slightly more by a couple of years

The notion of retirement and living out your golden years didn't even exist, especially the "golden" part.

To be sure, everyone, life was tough back then. Toilet paper wasn't invented until the 1850s. Toothbrushes weren't invented until the late 1700s. Life was short and generally tough, except for the very wealthy.

1st Industrial Revolution

The 1st industrial revolution began in 1750s-ish and lasted until the 1840s-ish. Historians love to argue the exact time frame but it's generally accepted to have been from the mid 18th century to the mid 19th century.

One of the major catalysts for the first industrial revolution was the use of water and steam to power mechanized factories. The textile industry, the dominant industry at the time, led the way with water- and steam-powered cotton, wool, and linen spinning machines, power looms, and the cotton gin (separation of the seed from cotton). In this regard, we went from people/animal power to water/steam power. This led to unprecedented increases in worker productivity, often as much as fifty-fold for daily output.

Other important advancements during this period of time included chemical manufacturing, iron production, the steam engine locomotive, cement, glass making, and gas lighting.

People began moving off their small farms (which they didn't own) to the cities to get jobs in the factories. The notion of employment, as we know it today, emerged. Large industries (textiles, manufacturing, mining, transportation, etc.) with multiple large competing firms began to form.

Average life expectancy in the U.S.:
- Men – 41
- Women – 46

2nd Industrial Revolution

The 2nd industrial revolution began in the 1870s-ish and lasted until the beginning of World War I, around 1914.

The big catalysts for the 2nd industrial revolution were electricity, the formalization of power grids, the internal combustion engine, the telegraph, and the telephone. So, once again, we shifted the basis of mechanization power, from water and steam to electricity and the gas-powered internal combustion engine. This led to bigger, better, and faster mechanized factories and again tremendous increases in worker productivity.

The telegraph and telephone enabled communications over great distances in a very short period of time. We used power grids to provide electrification to cities. Iron and steel production processes became much better. Paper-making processes greatly improved, leading to advancements in general education (basic reading and writing). The use of ammonia in fertilizer led to unbelievable crop yields.

More and more people moved from rural farms to find employment in cities, which cropped up around every major manufacturing operation.

Average life expectancy in the U.S.:
- Men – 68
- Women – 70

3rd Industrial Revolution

The 3rd industrial revolution began in the 1950s-ish. I think it's safe to say we're still in it today but rapidly moving into the 4th industrial revolution.

This is the age in which you've lived your entire life, punctuated by computing and digital technologies, communications technologies, the Internet, smartphones, and so on. It's often referred to as the *digital revolution*, the *digital age*, and the *information age*. (You are referred to as a *digital native*, while people in my generation are referred to as *digital immigrants*.)

In the 2nd industrial revolution, the biggest and most well-known companies in the world were those that could produce/sell the most physical goods. It included the likes of Standard Oil, General Motors, Ford, US Steel, Sears/Roebuck, General Electric, Bethlehem Steel, International Harvester, Borden, and Goodyear Tire & Rubber. [2]

As we near the end of the 3rd industrial revolution, today is quite different. The biggest companies in the world (using early 2022 market capitalization or market cap) are [3]:

1. Apple
2. Microsoft
3. Alphabet (Google)
4. Saudi Aramco (the only oil company in the top 10)
5. Amazon
6. Meta (Facebook)
7. Tesla
8. TSMC (Taiwan Semi-Conductor)
9. Berkshire Hathaway (investment company led by Warren Buffet)
10. Nvidia (U.S.-based computer chip maker)

Eight of the top 10 largest companies in the world are all tech companies. (Don't think for a moment that Amazon isn't a tech company. Its most profitable business segment is Amazon Web Services. Not familiar with AWS… read the Cloud Computing section in Chapter 11 on Infrastructure Technologies.)

Within the 3rd industrial revolution, we had some
interesting sub-revolutions.

- 3.1 Personal Computing, starting in the late 1970s… home
  computers, desktop computers, microcomputers, laptops, etc
- 3.2 Electronic Commerce, starting in the 1990s
- 3.3 SoLoMo (social-local-mobile), also starting in the 1990s and
  dominating the first 2 decades of the 21st century

Average life expectancy in the U.S.:

- Men – 75
- Women – 80

4th Industrial Revolution

And, finally, we have arrived, where we are today. (Put your seatbacks
and tray tables in their upright and locked positions. We are landing.)

It will be interesting to read about the 4th industrial revolution 20 years
from now. Did it start in 2105? Probably not. 2020? Perhaps. 2025? We
will most certainly be in it by then.

Regardless, the 4th industrial revolution is your present and your
future. You and your generation are in control. Historians 20 years from
now will write about how you and your generation shaped the 4th
industrial revolution.

As Captain Kirk, Picard, Sisko, Janeway, or Archer would say on *Star
Trek*, "You have the conn."

At the core of the 4th industrial revolution will be the following primary
technology drivers (see Figure 0.1 on the next page):

1. The Internet of (All) Things
2. Cryptocurrency & Blockchain
3. Artificial Intelligence
4. Extended Reality
5. 3D Printing
6. Autonomous Vehicles
7. Drones

Figure 0.1 **The 4ᵗʰ Industrial Revolution**

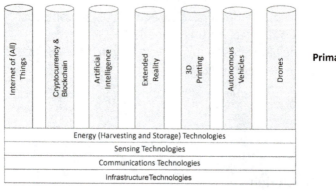

Those technologies will be supported and enabled by existing and advances in energy (harvesting and storage) technologies, sensing technologies (hearing, seeing, feeling, and smelling), communications technologies, and infrastructure technologies like quantum computing, cloud computing, and nanotechnologies.

And, we'll see applications and advancements of the 7 core 4ᵗʰ industrial revolution technologies in/with things like geo-engineering, bio-technology, neuro-technology, digital twins, and much more.

Most importantly, all those technologies are the foundation on which you will build unbelievable innovations. Those innovations will touch, impact, and disrupt every aspect of daily life.

No stone will go unturned.

See, I told you, a fast read. And please do understand that the history lesson on the last couple of pages was the equivalent of skipping a stone across a really, really, really big ocean and the stone only hitting the water a few times.

The history lesson continues. (Sigh.)

▲ ▲ ▲

The S-Curve of an Industrial Revolution

You can think of each industrial revolution as an s-curve (See Figure 0.2).

Figure 0.2 **Industrial Revolution S-Curve**

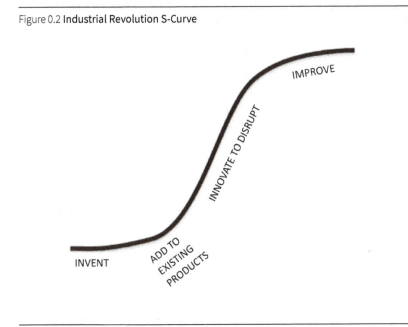

It all starts with a bunch of new technological inventions, like electricity, internal combustion engine, telephone, and telegraph of the 2nd industrial revolution. During the invention stage, market acceptance is low. At some inflection point, the rise over the run changes dramatically as the technologies get better, faster, cheaper, and more reliable, and as people start to innovatively figure out how to use the technological inventions to reshape existing products and services. Market acceptance sky rockets, the technologies continue to get better, faster, and cheaper, and society in general becomes more "comfortable" with new ways of doing things, not to mention that businesses figure out how to make money with their innovations. New products and services emerge based on the new technologies that weren't possible with previous technologies, leading to dramatic industry disruptions.

During the latter part of an industrial revolution, the speed and pace of innovation slows and people begin to focus on creating effectiveness (doing the right things) and efficiency (doing things right… in the least amount of time, at the lowest cost, etc.).

And then – suddenly – a bunch of new technological inventions emerge and the birth of a new industrial revolution begins.

Figure 0.3 captures the previous 275 or so years and our progression through the first 3 industrial revolutions and the beginning of the 4th. If you look to the upper right corner, you'll notice just the beginning of the s-curve of the 4th industrial revolution. Fortunate for you, we are just at the beginning of the 4th industrial revolution.

Figure 0.3 **275 Years in One Image**

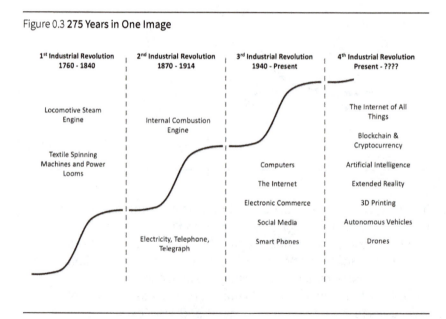

I believe, without a doubt, that the next 20 or so years will be the most innovative time in the history of the world. Your opportunities are simply unimaginable.

So, let's get to it.

The future is waiting for you, and you've got no time to lose.

Best of luck inventing the future.

Do a great job.

The world is counting on you.

▲ ▲ ▲

Innovation: Int or Ent?

Throughout my some 40 years of teaching at the collegiate level, I've spent most of it helping students learn the art of innovation. Within those interactions, I frequently come across students who say they don't want to own their own business, they don't want to be entrepreneurs.

Cool, I have no problem with that, but...

There are two kinds of innovators. There are *intrapreneurs* and there are *entrepreneurs*. Intrapreneurs want to go to work for an existing organization. Entrepreneurs want to build their own. And that's the only distinction, the only difference between the two.

If you want to strike out on your own, build something from nothing, suffer from and enjoy the roller coaster ride of the highs and lows of building a hopefully successful business, then you're an <u>entrepreneur</u>. If you want to work for an existing organization and help it offer new products and services, gain more market share, and create differentiation in the marketplace through innovation, then you're an <u>intrapreneur</u>.

Both groups are the same in that they question the status quo, ask why not instead of why, embrace new technologies, envision business not as usual, and – most importantly – understand that the world is changing and they must change with it.

Either way, you must be an innovator. If you're not moving forward, you're moving backward. If you're standing still, the competition is moving past you. If you're on top of the mountain, all of your competition is trying to take you down. If you're in first place, there is nowhere to go but in the direction of last place.

Is Amazon content with what it's doing, the status quo? Is Tesla? Is Meta?

Regardless of your choice – whether it be intrapreneurship or entrepreneurship, not-for-profit or for-profit, business or government or NGO – you must be an innovator.

We are, after all, in the *age of innovation* in the 4th industrial revolution.

▲ ▲ ▲

The Weed Whacker Justification

I like visual ques. My big visual que for this book has been a weed whacker. I even brought mine from my storage shed to my office so I could see it every day. (My wife eventually made me take it out of the house, as it smelled like gas. So, I printed a nice color photo of a weed whacker and taped it to my office wall.)

Why you ask? I'll answer.

You've heard the phrases "getting into the weeds" and TMI (too much information).

There are a lot of books about the 4th industrial revolution. Unfortunately, they take topics like quantum computing or blockchain and do a lot of hand waving and arm flapping but never really tell you what they are. They treat the technologies of the 4th industrial revolution like black boxes. Seemingly, you just apply the technologies and things *automagically* happen.

I'm a contrarian. I think you need to know something about how and why the technology works the way it does. If you understand the technology, then you can better envision and innovate possibilities.

But, there is a point at which we can dig too deep into the technology, that is, we can get too far into the weeds.

My hope is to introduce you to each technology without having to get out the old weed whacker. It's a delicate balance. How much of the technology do you really need to understand? In other words, when am I "getting in the weeds?" Throughout this book, I'll refer to the weed whacker. It's my que for "stop talking in so much detail."

Why Not, Not Why

When we're in the midst of "business as usual" and you propose a new product or service, you'll get the question of "why" from your boss or perhaps angel investors if you're striking out on your own. The question of why makes sense and you can address it with all sorts of market statistics, user reviews, comparisons to competitors, and the like.

But in the age of innovation when we're building entirely new things, *why* doesn't apply.

Instead, you should be answering the question, "Why not?"

Home computers, why not? Electronic commerce, why not? Social media, why not? TikTok. SnapChat. Netflix. Uber Eats and DoorDash. Metaverse. Venmo. Why not, why not,…

In retrospect, answering why seems obvious now for things like Uber Eats and DoorDash. But at their beginning, *why* didn't even make sense. But *why not* did. In fact, attempting to address the question of why would have meant that the innovations would never have become a reality. After all, who would have guessed that young people would be willing to pay a delivery fee for fast food. It didn't make sense.

In the innovation age of the 4th industrial revolution, seemingly wild and crazy innovations won't make sense and that's why we can't ask *why*. Instead, shrug your shoulders, ask *why not*, and set out to build a new future.

CHAPTER 1

The Internet Of (All) Things

A Water Bottle that Can Diagnose the Flu

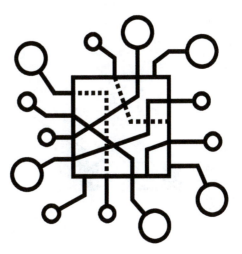

Are you smarter than a 5th grader? (No offense if you are a 5th grader.)

Question: Are there more stars in our universe or grains of sand throughout the world?

Scientists today estimate that there are 7.5 sextillion grains of sand, or 7,500,000,000,000,000,000. It's a big number. Our universe, however, contains an estimated 70 septillion stars, or 700,000,000,000,000,000,00 0,000. That means there are roughly 10,000 stars for every grain of sand.

So, you should get this question right: Are there more people in the world or more devices connected to the Internet? You guessed it... there are more devices connected to the Internet than the entire population of the planet by a factor of about 8 to 10, depending on when you read this.

As you can see in Figure 1.1 it was about 2007-ish when the number of devices connected to the Internet outnumbered the human population. [1]

Figure 1.1 **World Population Versus Internet Connected Devices**

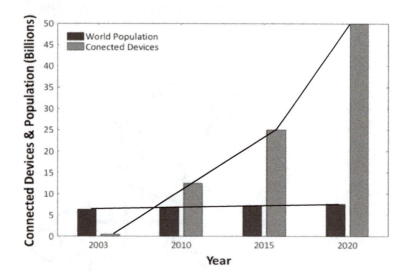

2020 had about 7 devices connected to the Internet for every person, and by 2030 Cisco estimates that there will be 500 billion devices connected to the Internet, about 60 devices per person assuming the world's population in 2030 is 8.5 billion. [2]

We aggregately refer to all those Internet-connected devices as the **Internet of Things (IoT)**, formally, a network of Internet-connected objects that collect, process, and exchange data.

▲ ▲ ▲

IoT Is Arriving in Stages

It's true that IoT already exists, but it's arriving in stages. We've got basic IoCT (Stage #1), we're deep in the middle of IoET (Stage #2), and we're just now moving into IoAT (Stage #3). Let's talk in more depth about the evolution of IoT (See Figure 1.2).

Figure 1.2 **Internet of (All) Things Evolution**

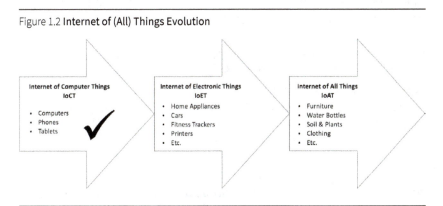

Stage #1: IoCT (Internet of Computer Things)

Stage #1, *the Internet of Computer Things* or *IoCT*, began in the 3rd industrial revolution with all your computers and computer-related things connected to the Internet, and communicating with each and sharing information, data, files, photos, etc. You can use your phone and share photos to your computer (AirDrop if you're in the Apple world, Nearby Share if you're in the Android world). You can use your laptop or tablet to locate your phone. You can easily and dynamically create, edit, and share files among your computer devices using services like iCloud, Dropbox, and Google Docs & Drive.

Seen it. Been there. Done that. IoCT is here and in full swing.

Stage #2: IoET (Internet of Electronic Things)

We are deep in the middle of stage #2, **the Internet of Electronic Things** or **IoET**, as we connect all electronic things to the Internet.

I have an HP printer connected to my home network. Using HP's service, every time I print something, the number of pages I print and the colors I use are sent to HP. HP uses that information to estimate when I need new printer cartridges. It sends them to me about a week in advance of when it estimates I'll run out of ink. I never run out of ink, and I never have too much. That's an IoET example focused on computer technology. But IoET does extend to all types of electronics.

Think here about household electronic devices such as washing machines, dryers, refrigerators, heating and air conditioning systems, garage doors, water sprinkler systems, stereos, and TVs.

Many of these are already a part of IoET. Consider Google Nest. It offers a wide range of Internet-connected home conveniences including Learning Thermostat, Doorbell, and Protect (smoke and carbon monoxide detector). Ring also offers a variety of IoT-based solutions for your home including indoor and outdoor security cameras and its signature video-based doorbell systems.

There are many other examples, many of which we consider in the **smart home** category (also called **home automation** or **domotics**), building automation systems and intelligence for the personal home. You can use an app to control your lights, garage door, and settings for your water sprinkler system. And, of course, that whole genre of be-as-lazy-as-you-want technologies like Amazon Alexa and Google Assistant can answer questions, control various parts of your home, and generally make you feel like you never have to leave your couch (or touch a remote control).

There's lots of other IoET applications as well for commercial buildings, transportation systems, weather alert systems, and so on. Many of these fall in the category of **smart cities**, technologically-based city environments that use a wealth of technologies to gather data and use that data to more efficiently deliver services related to such things as utilities, transportation, crime detection, medical care, waste removal, and snow plowing. Smart cities systems that apply IoT to improve the delivery, efficiency, and reliability of electricity are called **smart grids**. (Seems as if we started a whole *smart* craze when we came out with the smartphone.)

Industrial Internet of Things

All sorts of new terms and acronyms are popping up everywhere in the 4th industrial revolution space. One of those is the industrial Internet of Things.

The **industrial Internet of Things (IIoT)** refers to the use of IoT in commercial and industrial settings as opposed to consumer-oriented IoT like Google Nest and Ring. Smart cities use of IoT falls into the category of IIoT.

Major manufacturing operations are using IIoT to monitor manufacturing equipment. In this case, IoT sensors that measure vibration, for example, are attached to manufacturing equipment to detect early warning signals that equipment may be coming out of alignment and need maintenance and/or repair.

Both smart homes and smart cities rely heavily on the use of IoT.

We're here in the middle of the innovation aspect of stage #2, with still lots of interesting and innovative opportunities still to come.

Stage #3: IoAT (Internet of All Things)

We are just starting to dip our toes into stage #3, **the Internet of All Things** or **IoAT**, the connection of all things non-electronic to the Internet.

Let's fast forward to envision a possibility of the Internet of All Things. You get up one morning feeling a little groggy. Perhaps you didn't sleep well but something seems to be crawling up the back of your neck. You get ready and head off to school.

At some point, you take a drink from your water bottle and shortly thereafter get the text message to the right. It seems that your doctor's office received a communication from your water bottle that you may be coming down with the flu.

> Mickey, this is Dr. McGregor's office. It seems you're coming down with the flu. Please reply 1 to have our office call you.

Seem impossible? Not really and it may very well happen in stage #3, the Internet of All Things.

Clothing is becoming Internet-connected. We've already put sensors in shoes to measure your speed and distance travelled. Research is occurring to literally weave sensors into the fabric of clothing. [3] Someday, you'll start to wash a load of clothes and your washing machine will alert you to the fact that you left your car key fob in your pants pocket and are just about to wash it (the key fob, that is).

We can now connect sensors to living things like trees to monitor growth, determine when watering needs to occur, and detect the presence of diseases. (Many tree and shrub diseases emit chemicals, which we can detect with IoT *smelling* sensors. Learn more about smelling sensors in Chapter 9 on Sensing Technologies.)

We're also putting sensors in furniture. Pillows and beds can monitor your sleep patterns. [4] Intelligent chairs can give you a slight vibration when you've sat too long in the same position.

I love the water bottle innovation space from earlier. Imagine a water bottle that can track how much you've drank throughout the day and tell you when your hot coffee is losing its heat. You'll be able to combine your water bottle data with data from your home exercise system to get a more complete picture of your overall health and habits. And I do believe that sensors in water bottles can and will send information to your doctor's office for health analysis.

We can even 3D print sensors on to human skin. [5] These types of sensors provide real-time biometric data to enable better health care.

I think this is where you play, in stage #3 the Internet of All Things. We've just barely seen the tip of the iceberg in the IoAT space. Many, many wonderful innovations to come, just waiting for you and your free-thinking group of friends to bring them to market.

▲ ▲ ▲

IOT BY THE NUMBERS

Consumer spending on smart home devices is expected to double from 2020 ($44 billion) to 2025 ($88 billion). [6]

How IoT Works: From Sensor to Application

An IoT system has 4 layers. Consider Waze, the popular navigational app. (Yes, Waze is an IoT application.) Seems a bit weird, but read the table below from bottom to top.

IoT Layer	Waze
Layer #4: Application	Takes all that information, applies some algorithms and mapping functions, and shows you shortest routes, locations of traffic jams, location of emergency vehicles, etc.
Layer #3: Data Management	Aggregates and organizes the information from potentially millions of current active users by location.
Layer #2: Communication	Sends your information to the cloud via broadband cellular network (hopefully 5G).
Layer #1: Sensor	Uses your GPS to determine things like your speed, direction of travel, location, etc.

A few noteworthy elaborations.

Firstly, in the communication layer there are many options. If you're building an IoT application for use in or around your home, you'll mostly likely be using WiFi and perhaps Bluetooth, RFID, and/or NFC. Waze relies on broadband cellular (5G, etc.) because that's what your phone uses when you're on the road. Again, many options here and I would encourage you read Chapter 10 on Communications Technologies, especially if you're unfamiliar with things like RFID and NFC.

Secondly, is the notion of security. Most IoT systems fall in the category of *lean tech*, meaning that the system (usually because of its size) has only the minimum software and hardware necessary for it to execute its specific function or task. And, one of the first pieces of software to be omitted is security, unfortunately. The thinking is simple: You don't need security software to run your water sprinklers based on soil moisture level.

But the absence of security software can lead to hacks and breaches. In 2013 for example, hackers penetrated an HVAC company that wirelessly monitored retailer Target's environmental control systems. Using that wireless gateway to the heating and air conditioning systems, the hackers were able to access millions of customer records on Target's large computers. [7]

Of course you may think that hacking into your water sprinkler system won't do anyone any good. Think again. If someone can get into that IoT system, then they have made their way into your home's WiFi. That's trouble.

▲ ▲ ▲

IOT BY THE NUMBERS

Worldwide spending for all of IoT is expected to exceed $1.1 trillion by 2023. [8]

There's a Sensor for Everything

A **sensor** is an electronic device that detects, measures, and provides data about a thing. That **thing** could be just about **anything**... your location, someone walking into a room, the temperature, the presence of smoke, the pH level in water, etc.

There are, quite literally, hundreds of different types of sensors used in IoT applications, way too many to completely cover here. So, I'll provide the following groupings of IoT sensors:

1. GPS (location), accelerometer (speed), gyroscope (direction)
2. Proximity/Ultrasonic (how close is an object), motion, infrared, tracking (edge detection and line following), vibration
3. Barometric pressure, gas pressure, weight (pressure)
4. Temperature and humidity
5. Water level, water flow, moisture
6. Magnetic detection and metal touch
7. Chemical, gas, carbon monoxide/dioxide
8. Smoke, flames
9. Light
10. Sound
11. Water quality (e.g., pH, chlorine-related)
12. Image recognition, color recognition
13. Biometric (includes voice recognition, fingerprint recognition, facial recognition, and occupancy counting... how many people are in a room)

Like I said, hundreds and hundreds of sensors. If you need to detect and measure something, there's an IoT sensor for it. Costs range from about $1 to $50 and beyond. For example, I bought a nice AI camera for about $50. It does image and facial recognition. For less than $20, I bought a fingerprint recognition sensor that can store, recognize, and validate up to 240 fingerprints.

Natural or "Sensed" Data

I like to think about the whole IoT space in terms of natural or *sensed* data, captured data in its natural, raw form at the point of origin. I could count the number of people in a room, or I could use an IoT sensor that "senses" when a person walks into the room. Add 1. It could also sense when a person leaves the room. Subtract 1. That's very easy to do with IoT.

I've built toilet-base wrap-around IoT applications that detect water (moisture) when the toilet overflows. I've buried other moisture sensors in my backyard that turn on the sprinklers when the soil moisture drops below a certain level.

I strapped a GPS unit to my daughter's bicycle helmet so I could monitor her whereabouts on a bike ride. I built an IoT application and put smelling sensors in plastic bowls to detect when food was going bad in the refrigerator.

Organizations are quickly moving to take advantage of capturing natural or sensed data. Amazon with its retail stores, for example, is rolling out a cashier-less and self-checkout-less shopping experience. You just walk in, get what you want, and walk out. Amazon uses the Amazon Go app on your phone, camera technologies with image recognition, and IoT sensors to capture sensed data about you (when you enter the store), what products you select (using the cameras and IoT sensors), and your Amazon account (stored on the Amazon Go app) to charge you for the products you select.

You should think critically about the Amazon example. Any retailer using such a system will know how much time you spent in the store, the order in which you traversed through the store and selected products, what products you looked at but chose not to buy, and so on.

Capturing natural or sensed data yields so much more information than typical transaction data. Think about grocery shopping. When you check out, all your items get scanned. That's transaction data. But, the grocery store doesn't know which aisles you chose or how you moved through the store, in what order you selected your groceries, or what you pulled off the shelf, chose not to buy, and placed back on the shelf. That sort of sensed data would be invaluable for a grocery store.

We can capture natural or sensed data on the web very easily. Think about Netflix. It captures which movies you look at but choose not to watch. It captures movies for which you watch the preview but choose not to watch. That type of information is important to Netflix as it builds a profile of you and your movie habits.

Knowing what people choose not to buy is just as important – if not more important – than knowing what people do buy.

Tile, SmartTag, and AirTag as Examples of IoT Applications

We most often think of these as key finders, because keys are the most often lost item in the home. But you can attach Tile-like devices to just about anything… remote controls, wallets, gloves, shoes, etc.

The better ones of the market include:
- Tile (Tile)
- SmartTag (Samsung Galaxy)
- AirTag (Apple)
- Chipolo (Chipolo)

As an IoT application, each of these includes some sort of short-range communications technology to "find" or locate what you've lost. You can easily build your own version by incorporating something like Bluetooth.

▲ ▲ ▲

IOT BY THE NUMBERS

By 2025, the projected number of IoT device connections will be 3 times that of devices not connected via IoT. [9]

The Ease of the IoT Build Process

You can definitely build IoT applications, inexpensively and quickly. And, it's not that difficult to learn.

I love the IoT space because of the potential is has for unbelievable disruption and life-enhancing advancement. I also love it because it's a very easy space in which you can build. For about $15 I built a home detector for smoke and carbon monoxide. It took about 20 minutes. Of course, it was ugly and didn't have a nice enclosed unit that mounted easily to the ceiling. I also had to plug it into a wall outlet or use a D battery (which lasted about a day before needing a recharge). But, I built what Eric Ries calls a minimum viable product (MVP) in about 20 minutes for $15. (See Special Insert #2 titled *People Smarter than Me* for more on minimum viable products and Eric Ries. His book is a must read.)

Do a quick search on Amazon for IoT kits and you'll come across the likes of OSOYOO, Freenove, UCTRONICS, REXQualis, and ELEGOO. ELEGOO is my personal favorite, probably only because that's the one I've been building IoT applications with for several years.

The last time I bought an ELEGOO Mega R3 Project Kit it cost me about $60. It included sound, ultrasonic, water level, and other sensors as well as digital displays, a buzzer, a joystick module, remote control, a keypad, and lots of other cool stuff. For another $40 or so, I bought a 37-sensor upgrade kit.

You can also wonder around Amazon and find any type of other sensors that you might need for your innovation. You're now ready to build. Using one of these IoT kits, you connect everything by way of a breadboard, so no soldering is necessary. That's nice because you can easily dismantle your system and reuse the components to build a new one.

As well, you can incorporate communications technologies such as Bluetooth, NFC, or WiFi connectivity.

If you really think you have a great innovation and want to go to market, you'll need to refine your MVP until it works exactly the way you want. (By all means, do a lot of market testing with your MVP.) Then, you'll need to contact a printed circuit board (PCB) manufacturer and contract that organization to build your final product. That was the short explanation of what can be a very long process of diagramming your requirements, reviewing prototypes, doing quality control testing, ad nauseam.

▲ ▲ ▲

Edge Computing and IoT

Edge computing is a distributed computing model that moves the storing and processing of data closer to the where the data is captured or sensed. So, IoT can be an implementation of the edge computing model. The notion is to shorten the distance – both physical distance and time – between layers #1 and layers #3 and #4.

Again, think about Waze. Waze is definitely not about edge computing. When data is captured about your driving (location, speed, etc.), it is sent to the cloud where that data is stored, organized, and processed. The results are then sent back to your device in the form of maps, driving recommendations, etc. So, the distance between the capturing of the data and the storing and processing of the data is significant.

Let's go back to my water sprinkler IoT system. The sensors are in the ground around my house. The sprinkler control box is in the garage. I built another control unit and mounted it the garage wall. That control unit takes in the sensor data and instructs the sprinkler control box appropriately. That's edge computing because of the close proximity of the sensors and the processing of the data. Alternatively, my water company could develop such a system. All the sensor data would have to be sent back to the water company, which would analyze and send back instructions to my sprinkler control box. That's not edge computing.

Okay, now that you know the basics of IoT, let's talk about the future.

IMAGINE FORWARD

Everything Will Be Connected

With the continued miniaturization of technology, increases in functionality, and decreases in price, it's hard to imagine what won't be connected to the Internet via IoT.

Water bottles, backpacks, bicycles, clothing, furniture… those all begin to make obvious sense and the data we can sense from them can be used in many innovative ways.

Rugs in your house, a door, a doorstop, your lawnmower, even your toothbrush… at first glance those seem a bit less obvious for an IoT application. But simply stop and think innovatively. A toothbrush that can detect cavities and diseases, monitor brushing effectiveness, tell you when to spit. We'll see them someday. (Oops, Oral-B already has one and it includes artificial intelligence. [10])

Security Will Be a Huge Concern

I cannot stress enough how important security is as a consideration in the IoT space. People have already proven that they can take control of an autonomous vehicle through hacking. [8] To be sure, IoT is an integral technology for autonomous vehicles. I told you about the Target hack that occurred via the HVAC IoT system.

Even in the development of lean tech, we cannot sacrifice security.

The Death of the House Key

With everything connected and our ability to sense things with technology, we will see a significant shift in physical security toward the use of biometrics. Think about security in terms of:
1. What you have, a key or an identification card of some sort such as your school ID card.

2. What you know, a password.
3. Who you are, a unique biometric of you such as your fingerprint.

The first two levels are easy to steal and duplicate; the third not so much. We already have the use of biometrics on phones, computers, and other personal devices, and, to a certain extent, automobiles.

That trend will certainly continue and start to include things like access to your gym locker, bicycle lock, dorm building, apartment, and home.

You Don't Make a Pig Fatter Just by Weighing It

Good words of wisdom from the farm.

Gathering data does you no good if you don't use that data to take action or make better decisions.

If you want your IoT innovation to be successful, you have to find value in the sensed data.

Add Just a Pinch of...

So, how do you get started innovating in the IoT space?

If you recall our s-curve discussion, the application of new inventions starts by adding those inventions to existing products and services. It's an innovation technique I call, "Add Just a Pinch of..."

Take any existing product – a toothbrush, a water bottle, whatever. Now, simply add some IoT to it. What do you need to *sense*? What IoT sensors are required? What sensed data would you capture? How would you use that data? Now, build it. If you do so for a toothbrush, you just have to understand that your minimum viable product (MVP) will be ugly, ugly. You'll probably have to hold the technology in one hand while you brush with the other. But the point is to prove that the technology works, regardless of how ugly it may be. This is called a *proof-of-concept prototype*. It doesn't have to be pretty; it just has to work.

Try adding just a pinch of IoT to the following products. For each, identify what data will you sense and why, i.e., don't just weigh the pig.

- School backpack
- Running shoes
- Tennis ball
- Rain gutter
- Mountain bike
- Dog food and water bowl
- Toilet

As an alternative, you can focus on a specific type of IoT sensor and start to imagine possibilities of when/where it could be used. Take, for example, a simple pressure sensor that measures the presence of something by noting a change in weight. Can you envision possibilities for when this would be helpful for beds, car safety seats for children, lawn furniture, book cases, etc.?

As a final alternative, pick a big hairy world problem that you're passionate about. How can you use IoT to address that problem?

▲ ▲ ▲

IOT BY THE NUMBERS

Industrial (non-consumer oriented) spending on IoT is expected to be $110 billion in 2025, up over 23% from 2022 industrial spending levels. [11]

CHAPTER 2

Cryptocurrency And Blockchain

$250 Million for a Pizza

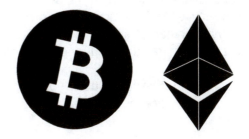

Time for another question: In 2010, someone bought the 2 most expensive pizzas ever sold. How much did those pizzas cost?

A. $1,000
B. $10,000
C. $100,000
D. $1,000,000
E. Upwards of a half-billion dollars

Sad story, as it turns out. (BTW, the answer is E.) In 2010, Laszlo Hanyecz had accumulated 10,000 coins of something called Bitcoin, a fanciful outlandish digital version of currency that wasn't backed by any government. At the time, a single bitcoin was worth $0.004, not even a penny. So, Laszlo spent the equivalent of about $40 in 2010 on 2 pizzas. $40 for a couple of pizzas isn't outrageous by any stretch of the imagination. But, in today's dollars 10,000 bitcoin is worth ever so slightly more than $40. (chuckle)

What's the value of 10,000 bitcoin today? Look up the value of a single bitcoin and quickly multiply that number by 10,000 and you'll get the exact amount Laszlo spent on 2 pizzas. (As I'm writing this, a bitcoin is worth $42k, so about $420,000,000 for 2 pizzas.) Ouch.

Of all the 4th industrial revolution technologies, cryptocurrency – along with blockchain – may very well be the most disruptive. No doubt, new industries, businesses, and ~~millionaires~~ billionaires will emerge in this space. Many traditional businesses and perhaps entire industries will not survive the disruption. Governments will have to rethink everything about monetary policy. You'll get to decide if you want your paycheck deposited in traditional currency or crypto.

I can't wait to see how you and your generation innovate around cryptocurrency and blockchain.

Cryptocurrency and blockchain are as well perhaps the most dominating of the 4th industrial revolution technologies in terms of their coverage in the popular press.

- In 2021, Aaron Rogers, the quarterback for the Green Bay Packers, decided to take part of his annual pay in Bitcoin and also give away $1 million in Bitcoin to fans. [1]
- In 2017, the Navajo Nation made a big push into mining Bitcoin. [2]
- In 2021, the newly elected mayor of New York City (Eric Adams) took his first three paychecks in Bitcoin. [3]
- In 2021, the city of Miami, FL announced it would distribute some of its Bitcoin yield to residents. [4]
- In 2021, Trevor Lawrence, the number 1 pick in the NFL draft, announced he would take his entire signing bonus (just over $22 million) in Bitcoin. [5]

I'll stop there, but you get the idea hopefully. You can do a quick search on cryptocurrency (and specific cryptos like Bitcoin and ETH) and blockchain and easily find hundreds – if not thousands – of recent articles. Those articles run the full spectrum, with some touting cryptocurrency and blockchain as the solution to most all our problems, and still others telling you cryptocurrency will never survive.

▲ ▲ ▲

CRYPTOCURRENCY AND BLOCKCHAIN BY THE NUMBERS

The blockchain market is expected to grow from $4.9 billion in 2021 to a predicted $67.4 billion by 2026. [6]

Cryptocurrency, Fiat Money, and Legal Tender

Before we get into the details (not the weeds, hopefully), let's talk briefly about money. You may be thinking you don't need an introduction to money, but let's make sure you're comfortable with a couple of terms.

The first is fiat. **Fiat money** is a currency that has been established as *money* by a government and through government regulation. The government controls the supply of fiat money, establishes policy, sets interest rates, and so on. That's the role of the Federal Reserve in the U.S., and, of course, our fiat money is the U.S. dollar. In Japan, it's the yen. For the European Union, it's the euro. You get the idea.

Next is legal tender. **Legal tender** is a form of money that a particular government's courts of law require to be accepted to settle debt. Again, in the U.S. the legal tender is the dollar. (I guess it's entirely possible for a government to create a fiat money that isn't legal tender, but that doesn't make much sense.) Interesting point here… there is no U.S. federal law that requires you or your business to accept folding cash and coins. The *form* of the legal tender that you choose to accept is entirely up to you. Those forms include personal checks, money orders, cashier's checks, wire transfers, credit cards, debit cards, and so on. You can also choose to accept alternative forms of payment including crypto like Bitcoin and ETH, foreign currency, and even fresh eggs and produce.

Okay, so let's now talk briefly about cryptocurrency. A **cryptocurrency (crypto)** is a currency in digital or electronic form, with no physical equivalent. Let's use Bitcoin for discussion purposes.

- That cool Bitcoin insignia you see everywhere is something completely made up.
- There isn't a pile of those lying around in a vault somewhere.
- Whenever new Bitcoin is released, there isn't a mint somewhere that produces the equivalent amount in physical form.

- You can't take your digital version of Bitcoin and go down to a bank somewhere and demand a physical Bitcoin. (I guess you could but expect to be laughed at.)

Crypto is *not* fiat money. We'll get more into this a bit later, but no government has created its own crypto and declared it to be fiat money for that particular country. (That statement was certainly true as of early 2022, but it will happen someday.) Legal tender is a bit different. In 2021, the country of El Salvador mandated Bitcoin be accepted as legal tender. [7] It will be interesting to see how that plays out.

One other quick note and then we'll move on… most countries have established laws and regulations regarding the use of cryptocurrency. Countries where certain cryptos are acceptable (but not required) for use include the U.S., those in the European Union, Australia, Canada, and many others. Some countries, though, have banned the use of cryptocurrencies within their borders. A few of those countries include China, Egypt, Iraq, and Morocco.

Okay, the hope and hype of cryptocurrency isn't solely that it's digital. (There is some allure to that, though. Consider in the U.S. that it costs a little over 2 cents to mint a penny and a little over 7 cents to mint a nickel. [8] Hmmm.) The supposed promise of cryptocurrency is the technology on which cryptocurrency is based, that is, blockchain, and the concept of a distributed ledger.

CRYPTOCURRENCY AND BLOCKCHAIN BY THE NUMBERS

As of March 2021, New York City reported that 261 businesses accepted crypto for in-store payment. Los Angeles followed closely with 217 businesses reporting the same. [9]

So, Who Invents Cryptocurrency?

As of early 2022, there were over 17,000 different cryptocurrencies in existence. [10] Again, because cryptocurrency is not fiat money (established by a government), quite literally any person can invent their own cryptocurrency. That may sound appealing but the technical complexities of doing so are enormous. And getting market acceptance is even more difficult.

So, who are the inventors of some of the more popular cryptos out today? They include:

- Bitcoin (BTC) – Satoshi Nakamoto (more on that fascinating story in a moment)
- Ether (ETH) – Vitalik Buterin
- Stellar (XLM) – Jed McCaleb
- Binance Coin (BNB) – Changpeng Zhao
- Cardano (ADA) – Charles Hoskinson
- Polkadot (DOT) – Gavin Wood
- Chainlink (LINK) – Sergey Nazarov
- Litcoin (LTC) – Charlie Lee
- Bitcoin Cash (BCH) – Craig Wright
- Tether (USDT) – Brock Pierce, Craig Sellars, and Reeve Collins

Satoshi Nakamoto gets the nod for really launching the *entire* cryptocurrency craze. Satoshi wrote a white paper in 2008 that described Bitcoin as a peer-to-peer electronic cash system. Interestingly enough, no one knows who Satoshi Nakamoto is. No kidding. We don't know nationality, if it's a man or woman, or if it's a single person or a group of people. (Craig Wright, the creator of Bitcoin Cash, claimed to be Satoshi in 2016 but later backed away from that claim.)

Before we dive headlong into cryptocurrency, we need first to talk about distributed ledger technology and blockchain.

Distributed Ledger Technology

In Figure 2.1, you can see the relationship between cryptocurrency, blockchain, and distributed ledger technology.

- All blockchain implementations are based on distributed ledger technology.
- But not all implementations of distributed ledger technology are blockchain.
- All cryptocurrencies are based on blockchain.
- But not all implementations of blockchain are cryptocurrency.

Figure 2.1 **Distributed Ledger Technology, Blockchain, and Cryptocurrency**

The vast majority of our technology systems today are based on a *centralized ledger technology*, with one "master" set of records stored in a centralized database, or ledger, and one central authority/organization that oversees the quality, updating, and use of the centralized database/ledger.

Your school has a centralized ledger for your classes and grades. The software your school runs provides all the rules for updating and use of that information. SeatGeek likewise has a centralized ledger of concerts, events, tickets, customers, and so on with accompanying software that controls the updating and use of that information. The U.S. banking system is based on the central ledger concept. That's why, when you deposit a personal check from a relative, it takes 2 to 5 days for the check to clear and for you to have access to the money. That check and the movement of money have to be processed centrally. Your bank has no way of knowing if the checking account of your relative (at another bank) has sufficient funds to cover the check.

The IRS, Amazon, TikTok… you get the idea. For almost every environment, there is a "master" set of information and a single authority with complete control over that information.

On the other hand, a ***distributed ledger technology*** is an environment in which all nodes on a distributed network have a copy of the master information and must aggregately agree on and approve the use and updating of the information. Thus, each node has a complete and up-to-date copy of the master information and the software for managing the information. As such, the master information (i.e., ledger) is <u>distributed</u> among everyone. This eliminates the need for a single authority to oversee the information and its use, updating, etc. When the nodes agree to the authenticity of a new transaction, each node updates their respective version of the master information, thus ensuring that every node has an accurate and up-to-date copy of the master information.

▲ ▲ ▲

CRYPTOCURRENCY AND BLOCKCHAIN BY THE NUMBERS

In early 2022, Bitcoin accounted for 41.9% of the total cryptocurrency market, while ETH accounted for 18.3%. The other 17,000+ cryptos accounted for roughly 40% of the market. [11]

Blockchain

Blockchain is an implementation of distributed ledger technology in which information is stored in the form of **blocks**, with each block holding multiple records or transactions, and with the blocks linked or **chained** together via a unique key or hash (See Figure 2.2). Again, all **blockchain nodes** (or simply **nodes**, the technical term for players and participants who verify transactions, maintain the integrity of the network, and keep an up-to-date copy of the ledger) must agree on the updates to and use of blockchain-stored information. And, all nodes have an up-to-date and accurate copy of the blockchain-stored information.

Figure 2.2 **Blockchain**

Let's consider a simple example. You and your friends have decided to start your own country but you don't want to use a centralized banking authority like that of the U.S. Federal Reserve. Instead, you agree that everyone (each person will be a node) will have to approve each financial transaction and that everyone will keep a copy (distributed) of all the financial transactions.

You decide to pay a friend for a product she makes. So, you tell all the nodes/people/citizens that you want to send her $10. Everyone gets word of the proposed transaction, validates that she and you are citizens, and that you have $10 in your account to send. All nodes approve the transaction and then the transaction gets executed, with each citizen updating his/her copy of the financial database/

ledger. Now, you have $10 less and she has $10 more in your respective accounts.

Sounds easy enough, but the technical implementation of a blockchain isn't.

How Blockchain Works

As I stated before, blockchain is an implementation of distributed ledger technology in which information is stored in the form of blocks, with the blocks linked or chained together via a unique key or hash. Each node has an up-to-date copy of the blockchain.

For purposes of illustration, let's assume that the entire U.S. banking and financial systems are now using blockchain. Every bank, savings and loan, credit union, stock brokerage firm, etc. are on the same blockchain system. Each of these is a node, has an up-to-date and accurate copy of the blockchain, and must approve all transactions.

> (Note here: In my previous example of your starting a new country, I talked about each citizen being a node. It was purely hypothetical. If and when our banking and financial systems move to blockchain, individuals like you and I won't be nodes. That is, we won't approve every single transaction. That will be the responsibility of nodes and miners, which may include banks, S&Ls, credit unions, brokerage firms, etc. These institutions don't have to be miners, but could be. More on miners a bit later.)

So, you write a personal check to a friend. She takes it to her bank to deposit it into her account. Her bank would send out the transaction information to all the other nodes (banks, S&Ls, miners, etc.) who would determine if the transaction was valid:

- Does your bank exist?
- Do you have an account with that account number with your bank?
- Do you have enough money in your account to cover the check?
- Does your friend's bank exist?
- Does your friend have a valid account with her bank?

If all the nodes run through the checklist above (by searching back through the blockchain to, for example, determine if you have opened an account with your bank and have the right account number) and give it the "thumbs up," the transaction gets executed and all the nodes write the detail of the transaction into their distributed ledger copy of the blockchain.

This continues to happen for all financial transactions, until a block gets full. Blocks do have limited space, determined mostly by the size of storage required for each transaction. Once a block is full, a unique hash or key is built that uniquely describes the contents of the block. It is added to the end of the block. That block is then written permanently by all nodes to their respective distributed copy of the blockchain. And every node starts a new block, with the beginning entry into the new block being the hash or key of the previous block. That block-ending hash or key becoming the block-starting entry for the next block is what <u>chains</u> the blocks together.

And please do understand that the process I described above is completely handled by software, no human intervention other than the teller at your friend's bank scanning in your check. And the process would be extremely fast, perhaps validating your check and moving the money into your friend's account in a matter of a few hundredths of a second, probably much faster.

Block Hashes and Keys

In blockchain, a **block hash or key** is a unique 256-bit 64-character series that uniquely identifies the contents of the block. These work on a simple trap-door principle… you can work forward starting with the block content to derive a unique hash or key; but it's impossible to reverse engineer and start with the hash or key and work backward to determine the block content that was used to derive the hash or key.

(Do I need to fire up the weed whacker?)

As an example, let's assume we have a completed block that has only 2 transactions:
- Arielle sent $100 to Francesca
- Francesca sent $100 to William

Using those 2 transactions, the derived 256-bit 64-character hash or key would be:

- f969f3239f5e7e07860cce91b3e32b3c479419dbe4cf5d1ebe0c-9e445d44b169

Again, this would be ending hash for that block and the starting entry for the next block, creating the chain.

Now, let's suppose that a hacker gets into the blockchain and changes the second transaction to reflect that Francesca sent $1000, instead of $100, to William. The new 256-bit 64-character series would be:

- 7136fc8454c7b0f60acf52a6689b70f930fc665a30b31567a229b-d6aced7e620

They are remarkably different. This is a nice security feature associated with using blockchain. If a hacker can ever get into a blockchain and change a transaction, the hash or key changes and no longer matches the starting hash for the next block, thus breaking the chain. This causes a system error. Recall also that every node on the network has an up-to-date copy of the blockchain. So, if you were a node and got hacked, your copy of the blockchain would not match the copy of the blockchain held by other nodes, creating another system error.

(If you'd like to play around with a 256-bit 64-character key, visit https://passwordsgenerator.net/sha256-hash-generator/.)

Non-Cryptocurrency Implementations of Blockchain

Our discussions of blockchain to this point have focused on cryptocurrency. And blockchain is an essential technology for cryptocurrency; we can't have cryptocurrency without blockchain.

But not all blockchain implementations are cryptocurrency. Indeed, blockchain is applicable to just about any and every use of technology to store, manage, and verify the integrity of transactions and the stored information.

- **Voting** – The USPS has filed for a patent using blockchain to create, store, and verify voter registration and voting records. [12] Within in this system, each citizen will have a unique voter ID and their votes will be stored on the blockchain, with election officials participating in the verification of the votes.

- **Diamond Tracking** – In 2020, the DeBeers Group announced that it successfully piloted a blockchain-based system for tracking diamonds. [13] The system tracked each diamond from initial mining of the diamond all the way through retail sale.
- **Supply Chain Management** – Walmart is using blockchain to manage its supply chain management process. [14] Using blockchain, Walmart has created a system to automate its invoices from and payments to its 70 third-party freight providers in Canada.
- **Real Estate** – The full gambit of the real estate industry – residential, commercial, vacation rental homes, and even parking spots – is poised to adopt blockchain technology. [15] On the residential front, for example, blockchain can be used to track home ownership, mortgages, lawsuits for payments for work done, county inspections, and much more.
- **Healthcare** – The healthcare industry similarly is exploring the use of blockchain for a variety of applications. [16] Everything from patient medical records, insurance payments, and the tracking of disease outbreaks can be stored and managed on a blockchain.
- **Financial Services** – Many of our illustrations of blockchain have been around financial services, and there are many more. [17] Some of those include traditional banking services like checking accounts, stock market transactions, identity verification, and credit reporting.

Blockchain may hold great promise for making basic financial services available to the almost 1.7 billion adults worldwide who are **un-banked**, meaning that they do not have an account with a financial institution or a mobile money provider. [18] Many of these un-banked people cannot afford a traditional account because of the fees. **Decentralized finance (DeFi)** is a type of emerging financial technology based on distributed ledger technologies (i.e., blockchain) that may very well eliminate the fees associated with traditional financial account services like checking and savings. DeFi falls within the broader category of **fintech**, the integration of technologies – like blockchain – that improves the delivery, efficiency, and cost of basic financial services.

The list of promising applications of blockchain goes on and on. Pick any industry, pick any information- or transaction-intensive environment... they can all benefit from blockchain.

Smart Contracts and Dapps

A **smart contract** is a piece of software that runs on a blockchain and automates processes, tasks, and the movement of money as certain conditions are met. Within a smart contract, there are a number of nodes that must verify that those conditions have been met.

Suppose, for example, that you want to build a home. You would place money in an escrow account, which the smart contract would manage on a blockchain. As your general contractor and subcontractors complete certain aspects of your home construction, the smart contract would require that the appropriate nodes verify the work. Once all nodes verify the work, payment is automatically deducted from your escrow and sent to the appropriate contractor.

As an illustration, you would have a subcontractor responsible for the electrical work on your home. After that subcontractor completes the work, it would have to be verified and approved by the general contractor, the county inspector, and you as the owner. Once those three nodes approve the work, the electrical subcontractor is immediately paid by the smart contract using your escrow monies.

Rinse and repeat for plumbing, framing, roofing, carpeting and flooring, painting...

Decentralized apps (Dapps) are also software applications that run on a blockchain. There are some subtle and important differences between smart contracts and Dapps (and a lot of overlap), but I'm not going to delve into them here. (I'd need the weed whacker for that discussion.)

Dapps include the likes of: [19]
- LBank – a crypto asset trading platform
- PokerKing – online gambling
- CryptoKitties – trading and gaming platform
- Brave – web browser

- EOS Dynasty – role-playing, player versus player blockchain-based game
- TRACEDonate – for tracking how your money gets used when you make a charitable donation
- Circulor and Chainyard – supply chain management

Decentralized Autonomous Organizations

I love the simplicity of the definition of a DAO provided by Cooper Turley. According to Cooper, a ***decentralized autonomous organization (DAO)*** is an Internet community with a shared bank account. [20] A DAO is basically a group of people who decide to pool their money for any number of purposes and establish a set of rules by which the group will operate.

Creating a DAO is rather like forming a partnership with several of your friends. You pool your money into a single account, and then you establish the rules by which your partnership will operate. What makes DAOs particularly appealing is that you automate your operating rules in the form of a smart contract on a blockchain.

Let's suppose your partnership wants to get into investing in rental real estate. You could establish automated rules around the approval of a purchase, for example that 100% of the partners must approve the purchase of a new piece of real estate. You could establish rules around the distribution of profits, for example that 20% of the profits will be held in escrow in case you need funds for emergency repairs and maintenance. You could establish rules around the sale of a piece of property, for example that 90% of the partners must approve the sale.

Again, all these rules would be automated on the blockchain. Every partner would receive notification of the pending execution of a rule and be provided the right to vote on its approval.

Obviously, DAOs are mainly prevalent in the blockchain and cryptocurrency worlds. A DAO needs software that enables the automation of rules, that is, smart contracts that run on a blockchain. And, as most DAOs have some sort of financial mission (investment, etc.), the money medium is most appropriately cryptocurrency because of its ease of use on blockchains and in smart contracts.

Some examples of well-known DAOs include: [21]
- PleasrDAO – invests in NFTs (discussed later) and other digital investments
- HerStory – funds projects by black women and non-binary artists
- Komorebi Collective – funds women and non-binary crypto founders
- Friends with Benefits – pay-to-enter exclusive social club
- MetaCartel Venture – invests in early-stage decentralized applications

▲ ▲ ▲

CRYPTOCURRENCY AND BLOCKCHAIN BY THE NUMBERS

In January 2021, a survey reported that 66% of U.S. consumers had not invested in cryptocurrency and were not interested in doing so. Only 5% responded that they had invested in cryptocurrency and liked it. [22]

Cryptocurrency

Okay, we've been talking a lot about cryptocurrency to illustrate distributed ledger technology and blockchain. Let's now take a deeper dive into that fascinating topic.

Public and Private Keys

As I've already stated, when you get crypto (a friend sends it to you, you buy some on an exchange, etc.), you don't get – and you can't get – a physical coin. All cryptocurrencies are completely digital. As well, all cryptocurrency use cryptography to secure, validate, and authenticate transactions. That's why you see "crypto" in front of the word currency.

Formally, the type of cryptography that all cryptos use is called **Public Key Cryptography (PKC)** or **Asymmetric Encryption**, a trap-door approach that makes use public keys and private keys to secure transactions.

A ***public key*** is a key on a blockchain that is publicly known and is used mainly for identification. In the crypto world, someone has to know your public key in order to send you cryptocurrency. A ***private key*** is something very private, and – without getting into the weeds – is an electronic key that gives you the sole and exclusive right to use your cryptocurrency. Think of a public key as something on a blockchain everyone can see and thus identify that you are the owner of cryptocurrency. Your private key is what gives you the right to use that cryptocurrency.

So, when you buy cryptocurrency (or a friend sends it to you), you receive a private key for it, giving you the exclusive right to use it, send it to someone else, trade it for another cryptocurrency, etc.

To keep your cryptocurrency safe, you need to securely store your private keys in a cryptocurrency wallet or on a cryptocurrency exchange.

Cryptocurrency Wallets and Exchanges

There is definitely a difference between a cryptocurrency exchange and a cryptocurrency wallet. A ***cryptocurrency exchange*** is a marketplace on which you can buy and sell crypto. Popular cryptocurrency exchanges include Coinbase, Exodus, Binance, Kraken, and Gemini. Think of a cryptocurrency exchange as your bank holding your bank accounts, and thus your money. All cryptocurrency exchanges offer wallet capabilities.

A ***cryptocurrency wallet*** is a software system that allows you to store and manage your cryptocurrency. Again, all exchanges offer wallet capabilities. The downside to most of these is that the exchange holds and manages your private keys to the cryptocurrency you own. Exchanges can be hacked and your private keys can be stolen. If so, you may no longer own that cryptocurrency because the hacker can use your stolen private key to move the crypto elsewhere. (It's gone.)

In 2021, Liquid, a Japan-based cryptocurrency exchange, was hacked with the hackers gaining access to and making off with almost $100 million in cryptocurrency. [23]

So, many experts will tell you that you need to store your cryptocurrency on a cryptocurrency wallet not offered as a web-based wallet by a cryptocurrency exchange. Popular hardware-based cryptocurrency wallets include Ledger, Trezor, and SafePal. These allow you to move your cryptocurrency (i.e., private keys) offline. Ledger, for example, is a USB-connected flash drive for your computer. When you move your cryptocurrency there and then unplug it, no one can access the private keys to your cryptocurrency. Ever.

Of course, if you happen to lose that USB flash drive, it's the same as losing your physical-money wallet or your money clip. Your money is gone. Bye-bye.

Stablecoins

Because cryptocurrencies like Bitcoin and ETH are not regulated by a particular government through monetary policy, they tend to be extremely volatile. On January 1 of 2021, Bitcoin was selling for just over $29,000 and ETH was selling for just over $1,200. At the end of

2021, Bitcoin was selling for a little over $47,000 and ETH was selling for a little over $3,800. In between, Bitcoin rose as high as $67,000, while ETH rose as high as $4,800. [24]

Because of this volatility and rather extreme change in value, it's practically impossible for any government to declare that something like Bitcoin or ETH to be its fiat money. (As well, since a government doesn't control the supply of a cryptocurrency, it cannot regulate it through policy.) To reduce the volatility of cryptocurrency and make it act more like traditional fiat money, people invented stablecoins.

A **stablecoin** is a type of cryptocurrency that is pegged or tied to an external asset like the U.S. dollar or gold. This helps stabilize the price of the stablecoin. Popular stablecoins include: [25]

- Tether (UDST) – the U.S. dollar
- Dai (DAI) – the U.S. dollar (loosely)
- Binance USD (BUSD) – the U.S. dollar
- TrueUSD (TUSD) – the U.S. dollar
- USD Coin (USDC) – the U.S. dollar
- TerraUSD (UST) – the U.S. dollar (loosely)
- Digix Gold Token (DGX) – the price of gold

You can buy any of these stablecoins, and many others, on popular cryptocurrency exchanges.

Of course, the biggest advantage of owning stablecoin is that $1 in Tether will still be worth $1 in the future. The purchasing power of that $1 may have declined because of inflation. But the same is true for your money in your bank account.

Cryptocurrency Mining

During the creation of a block, the nodes on the blockchain are responsible for validating transactions, creating the ending hash for a block, keeping an up-to-date copy of the blockchain, etc. Some of these nodes are referred to as miners. **Cryptocurrency mining (mining)** then is the process of verifying transactions in a block, creating the ending hash for a block, and maintaining the integrity of the network.

The question then becomes, "How do miners get paid for all that work?" (Great question if you were thinking that.)

Once a block is full, then the race is on to create the unique hash that describes all the information in a block. This involves solving a hugely complex mathematical problem using the information in a block to create a unique hash that uniquely describes the information in a block, such that even the slightest change to the block of information will cause the hash to be different.

When a miner is the first to solve that math problem, they send it out the hash to all other miners for verification. The other miners verify the hash, and then all miners/nodes write the block to their respective distributed copy of the blockchain.

The miner that solves the problem and builds that hash is rewarded monetarily. Using Bitcoin as an example, the miner who "wins the race" by solving the computational problem first is paid 6.25 bitcoin (as of early 2022). And, this is completely new bitcoin. That's how new bitcoin enters the market; new bitcoin is generated and entered into the system by paying the miner who wins the race. (You should most definitely read the side box).

Paying Bitcoin Miners

When Satoshi Nakamoto wrote the original paper on Bitcoin, he/she/they outlined the payment mechanism for mining. Satoshi also included the fact that there would only ever be 21 million bitcoin.

During the early years of Bitcoin, miners were paid at a rate of 50 bitcoin per block. That has been halved roughly every 4 years. As of 2022, Bitcoin miners are paid at a rate of 6.25 bitcoin per block.

6.25 bitcoin may seem like a lot of money, as it was $250,000 in early 2022. But (and it is a big but), the initial technology investment into all the hardware and software for mining

Bitcoin will easily be in the millions of dollars. (Seriously.) It takes a lot of computing horsepower and electricity to be a Bitcoin miner. And, there's no guarantee that you'll win the race. There are probably upwards of a million Bitcoin miners today worldwide.

Getting Started Investing in Crypto

It's pretty simple. I would recommend that you register for an account with a cryptocurrency exchange like Coinbase. Once you register (it's free), you'll have a web-based wallet and you can start to invest. Some things to keep in mind:

- Many exchanges require you to be at least 18 years of age.
- All exchanges have country limitations; some countries won't allow you to buy cryptocurrency.
- When you do sell, all exchanges will want tax information; you will get a 1099 because the U.S. government currently treats cryptocurrency as an investment.
- Look at the fee structures carefully. These differ from exchange to exchange.

Most importantly, and I'm very serious about this… *don't bet hurt money*. Treat your investment in cryptocurrency as purely speculative. If you're on a tight budget right now, cryptocurrency is a dangerous play because of its volatility. But, if you have a hundred bucks lying around, then find a few cryptos that interest you and spread your money among them… something about, "Don't put all your eggs in one basket." And, be prepared to just leave it alone. See how you're doing 6 months from now.

The Benefits of Blockchain-Based Cryptocurrency

So, now you have some idea of how blockchain-based cryptocurrency works. And blockchain is essential for cryptocurrency, and its many other applications. For purpose of our discussion of the benefits and advantages of blockchain-based cryptocurrency, let's shift our example to Ether, the second most popular crypto. Ether is often just referred to as ETH (long e-th).

No Double Spending: Let's assume you buy a single ETH coin. Every ETH coin (and fractional part of an ETH coin) has a unique ID, of sorts (i.e., the private key). Let's say the unique ID for your ETH coin is 1234. When you buy ETH 1234, your ownership of it is stored on every distributed copy of the blockchain on the network. If someone else were to try to spend your ETH coin (1234) or claim ownership to it, the nodes on the network would invalidate the transaction because you are the owner and not the other person.

So, no one can make a copy of your ETH coin. There can never be 2 ETH coins with the unique ID of 1234.

Immutability: **Immutability** is a characteristic of a blockchain that basically says once a transaction has been approved and recorded it cannot be changed or altered in any way. So, there are no erasers in the blockchain world. If, by accident, you enter an incorrect transaction, such as selling 2 widgets when you actually sold 20, you would have to submit another transaction to reflect the sale of the other 18. You cannot "erase" the 2 in the previously approved transaction and write in the 20. If you were to do this, the hash for that block would change and no longer match the beginning hash in the next block, resulting again in a system error.

Non-Fungible Tokens (NFTs)

A **non-fungible token (NFT)** is a blockchain-based token that is non-interchangeable and represents ownership. (I'm sure that's about as clear as mud.) Think of it in economic or financial terms. Traditional currency is fungible. Think about a $20 bill. It represents that you have $20 in value for purchasing something. And, you can exchange a $20 bill for 2 $10 bills. That exchange ratio has been formally defined by the U.S. government and includes all denominations of coins and paper money.

Non-fungible tokens, while they can be extremely valuable, are very different. Someday when you a buy a car, you may get an NFT representing ownership of that car, instead of the traditional car title.

The NFT will be stored on a blockchain and represent that you own that car. When you sell your car, you'll need to know the new owner's public key and you'll use that public key to send the new owner the NFT for the car and the private key that gives him/her the exclusive right to the car.

Likewise, there is no formal exchange ratio for NFTs. There's nothing out there that says your 2021 Toyota Corolla is worth exactly 2 2011 Honda Civics. If you wanted to "swap" your 2021 Corolla for 2 2011 Civics, you would have to sell your car (and representing token or NFT) to someone and in turn he/she would sell you the 2 2011 Civics (and representing tokens or NFTs).

What really became popular in 2021 was the use of NFTs to represent ownership of digital content like digital art, video clips, memes, virtual real estate in metaverse, trading cards, and the like. When I wrote this, the most expensive NFTs ever sold for digital content included: [26]

1. Pak's *The Merge* – $91.8 million
2. Beeple's *Everydays: The First 5000 Days* – $69.3 million
3. Pak and Julian Assange's *Clock* – $52.7 million
4. Beeple's *HUMAN ONE* – $28.985 million
5. CryptoPunk *#7523* – $11.75 million
6. CryptoPunk *#3100* – $7.67 million
7. CryptoPunk *#7804* – $7.6 million
8. Beeple's *Crossroad* – $6.6 million
9. XCopy's *A Coin for the Ferryman* – $6.034 million
10. Beeple's *Ocean Front* – $6 million

I know the numbers seem unbelievable, but they are true. All of those dwarf the high-profile sale of a video clip of a massive dunk by Lebron James. NBA Top Shot sold it for a little over $200,000.

This truly is the wild, wild west. In September 2021, Snoop Dogg announced he was selling 1,000 NFT passes for people who wanted to virtually party with him in Sandbox, a popular metaverse destination. [27] That same year, an investor paid $1.23 million to own the virtual land (via an NFT) next to Snoop Dogg in Sandbox. That's a lot of money to be Snoop Dogg's virtual next-door neighbor. [28]

You, too, can mint your own NFTs, for just about any digital content you care to create. If you do, you'll want to sell them on an NFT exchange. The more popular ones include: [29]

- OpenSea
- Axie Marketplace
- Larva Labs/CryptoPunks
- NBA Top Shot Marketplace (NBA-specific NFTs only)
- Rarible
- SuperRare
- Foundation
- Nifty Gateway
- Mintable
- Theta Drop

Many of those NFT exchanges have tools so you can build your own NFTs.

Okay, we need to wrap our time here on blockchain and cryptocurrency. Like all the other 4th industrial revolution technologies, I've only scratched the surface of blockchain and cryptocurrency. I would definitely encourage you to continue your learning in this space.

▲▲▲

IMAGINE FORWARD

Cryptofiat

In the next 3 to 5 years, a major country will announce that its fiat money is now a cryptocurrency. I think that first big country will be China. For the past several years, China has been piloting its own national cryptocurrency, called the digital yuan.

It's just a matter of time – and not much time – before China adopts the digital yuan as its fiat money.

Other countries will follow.

Information Storage and Updating Will Evolve to Blockchain

Way back in the 1970s and 1980s, information storage evolved from *files* to *databases*. In fact, the term database is ubiquitous in information storage, and simply refers to a collection of files. So, files didn't go away; they simply morphed into databases.

The same will be true for databases and blockchain. Databases aren't going away. But, they will morph into blockchain-based storages of information.

Recommendation: If you want to be in the IT field, take a course in blockchain. It's the future.

PayPal, Venmo, and Others Will Pivot to Cryptocurrency

Services like PayPal and Venmo are peer-to-peer financial networks. They enable you to easily move money – the U.S. dollar for example – to another person.

That *money* will become cryptocurrency. (PayPal already lets you buy and sell cryptocurrency.)

The Death of Folding Cash and Coins

This certainly seems inevitable. It costs a lot of money to mint folding cash and coins. And, when fiat money becomes cryptocurrency, folding cash and coins will become those seldom-seen oddities. It probably won't happen in my lifetime, but it will in yours.

▲ ▲ ▲

CHAPTER 3

Artificial Intelligence

A Cat Mistaken for Guacamole

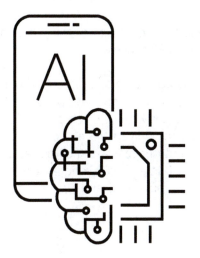

The movie became an instant cult classic… an AI (artificial intelligence) got so smart that it understood the concept of procreation, locked a woman inside her home, and impregnated her resulting in her having a computer baby. Sound familiar? Probably not. The movie was **Demon Seed**, and it came out in 1977, some 45 years ago.

Of course, you've watched untold numbers of more recent movies that depict AI… the **Terminator** series, the **Matrix** series, the **Alien** series, the **Iron Man** series (loved J.A.R.V.I.S.)… list goes on and on. It seems Hollywood has an unbelievable fascination with artificial intelligence. Of course, Hollywood produced those movies because we as a society are even more fascinated by artificial intelligence.

As an academic discipline and field of study, artificial intelligence has been around since the 1950s. Some 70 years later, a Harvard professor showed how Google's AI-based image recognition technology could be (easily) fooled into believing it was looking at guacamole when, in fact, it was an image of a cat. [1] Needless to say, we have a ways to go before artificial intelligence becomes as smart as us humans.

As a formal definition, let's say that **artificial intelligence (AI)** is a machine that mimics human intellectual tasks such as learning, problem solving, and social interaction. If you happen to search on artificial intelligence, I can guarantee that you won't find that exact definition. Instead, you'll find hundreds of different definitions. But, they all focus on the same thing, making machines seem like they have human intelligence.

Just a word of caution before we dive into artificial intelligence. Without a doubt, AI is the most complicated of all the 4th industrial revolution technologies we'll be discussing. Volumes have been written about the subject. Universities and colleges offer entire degree programs in AI, many at the masters and PhD level. I checked out the Wikipedia page for AI in early 2022 and it included over 450 explanatory notes, citations, references, and further readings.

Big and complicated field.

We're going to try to cover it in less than 25 pages, without using the weed whacker too many times. Please pardon my technical inaccuracies.

**AI BY
THE NUMBERS**

The global AI market is predicted to reach $997 billion by 2028, up from $93.5 billion in 2021. [2]

The AI Spectrum: From Single Task to Singularity

The spectrum of AI applications, research, and dreams goes from AI performing a single task to AI achieving *singularity*. Most commonly, AI is presented in terms of the 7 categories in Figure 3.1. But, these categories are not mutually exclusive, which makes this field even more confusing and complicated. I'll explain the overlaps after introducing each category to see if we can reduce the confusion and complication. No guarantees, but we'll try.

Figure 3.1 **The 7 Categories of Artificial intelligence**

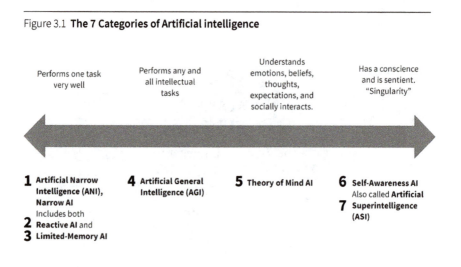

| Performs one task very well | Performs any and all intellectual tasks | Understands emotions, beliefs, thoughts, expectations, and socially interacts. | Has a conscience and is sentient. "Singularity" |

1 Artificial Narrow Intelligence (ANI), Narrow AI
Includes both
2 Reactive AI and
3 Limited-Memory AI

4 Artificial General Intelligence (AGI)

5 Theory of Mind AI

6 Self-Awareness AI Also called **Artificial**
7 **Superintelligence (ASI)**

Artificial Narrow Intelligence

On the far left is ***artificial narrow intelligence (ANI* or *narrow AI)***, an AI that performs one and <u>only</u> one task. Seems rather simple and mundane, but this is where almost all current successful implementations of AI fall, getting a machine to perform one intellectual task very well.

Artificial narrow intelligence includes the likes of Netflix's recommendation engine, car engine diagnostics, chatbots, virtual assistants, spam filters, image recognition, handwriting recognition, medical diagnosis, and even autonomous vehicles. Each of these performs a single task. Netflix's recommendation engine, for example, helps you hopefully find a movie that you'd like. It can't tell you which books to read or classes to take in school. It can only recommend what video you might like to watch. Period.

Reactive Artificial Intelligence

Reactive artificial intelligence is a subset of artificial narrow intelligence in which the AI is programmed to provide a predictable outcome based on the inputs it receives.

- The same set of inputs will always produce the same outcome.
- It will always respond to the same situation (inputs) in exactly the same way each and every time.
- It cannot learn to handle new inputs nor can it adapt itself to new situations.
- It cannot *learn*.

Let me give you an example of a reactive AI for determining what to do when approaching a traffic light at an intersection. You can build a reactive AI for this task in the form of rules as in the table below.

	RULE/QUESTION	YES	NO
1	Is the light red?	Go to Rule #2	Go to Rule #3
2	Do you have time to stop?	Stop	Accelerate through the intersection and hope you don't crash.
3	Is the light yellow?	Go to Rule #4	Proceed through the intersection.
4	Do you have time to stop?	Stop	Accelerate and proceed through the intersection.

Let's work through the example with the light being green. The AI would check the first rule, "Is the light red?" The answer is no, so it proceeds to rule #3. It checks that rule, "Is the light yellow?" The answer is no, so it would tell you to proceed through the intersection.

Similarly, if the light were red, the AI would check the first rule, get a "yes," and then proceed to rule #2. At rule #2, the AI would have to determine – based on your speed and proximity to the intersection – if you have time to stop. If the answer is yes, it would instruct you to stop. If the answer is no, then of course you may have a serious problem.

This type of AI tool is also called an expert system. An **expert system** uses a series of if-then rules to work through a set of inputs to determine what action to take. (Expert systems have been around for quite some time. I started programming expert systems way back in the 1980s using a language called LISP, which stands for **LIS**t **P**rocessor.)

In this simple example, you can see that a reactive AI will always arrive at exactly the same outcome based on 2 simple inputs, the color of the light and whether or not you have time to stop.

You can also see that this AI cannot take into account any other inputs, such as weather, the presence of pedestrians, a left-turn signal, what other vehicles are doing, the presence of emergency vehicles, and so on. It has been programmed to *react* to only 3 colors of a traffic light and whether or not you have time to stop.

If you wanted this AI to take into account the presence of emergency vehicles, for example, you would have to change the way the AI works by altering the if-then rules and adding a couple of more. It would look like the table on opposite page.

Examples of reactive AI include Deep Blue (IBM's AI-based supercomputer that beat Garry Kasparov, the world champion chess player, in 1997), spam filters for our email, car engine diagnostic software, Google's AlphaGo, and Netflix's recommendation engine (and most all the recommendation engines out there).

	RULE/QUESTION	YES	NO
1	Is an emergency vehicle in the area?	Go to Rule #6	Go to Rule #1
2	Is the light red?	Go to Rule #3	Go to Rule #4
3	Do you have time to stop?	Stop	Accelerate through the intersection and hope you don't crash.
4	Is the light yellow?	Go to Rule #5	Proceed through the intersection.
5	Do you have time to stop?	Stop	Accelerate and proceed through the intersection.
6	Is the emergency vehicle travelling parallel to you?	Pull over at least one lane and wait for the emergency vehicle to pass.	Stop where you are and wait for the emergency vehicle to pass.

Netflix's recommendation engine uses several mathematical and statistical models to make recommendations for you. The more movies you watch, the more reviews you provide, and the more searches you perform on Netflix, the better the system gets at making recommendations for you. That may seem like "learning," but it isn't. It simply has more data at its disposal to make a better recommendation.

Limited-Memory Artificial Intelligence

Limited-memory artificial intelligence is a subset of artificial narrow intelligence in which the AI learns from historical/past data to make better decisions. So, the more limited-memory AI is used, continually learning, the better it gets at making its decision.

Autonomous vehicles definitely fall into this category of AI. Previous to your buying your autonomous vehicle, its AI has been trained in a simulated environment. It's driven perhaps millions of miles on a simulator to learn how to drive, react to different road conditions, handle traffic congestion, and so on. As you drive it (or allow it to drive itself), it continues to learn, getting better at the task of driving.

In this way, limited-memory AI is actually adapting itself to get better at the task of driving. This is analogous to how you drive over time, continually learning and adapting your driving technique to get better.

Limited-memory AI is the dominant category of AI. As it is so dominant, we'll come back to this category of AI in a moment.

Artificial General Intelligence

Artificial general intelligence (AGI) is an AI that can take what it knows and has learned for one situation and apply it to a new situation. These AI will function completely like a person, intellectually. They will be able to build competencies in situations without being explicitly programmed for them. They will be able to generalize across domains, adapting their current knowledge to new situations. When using the term "new situation," I'm not referring to good-weather driving versus driving in the rain. I'm referring to an AI learning to drive a car and then being able to adapt that knowledge and literally teach itself how to drive a motorcycle or speedboat.

This is an important goal in the field of AI, and we are not there yet. Indeed, many people don't believe we'll ever be able to create a general AI that can adapt to new problems and situations, without some sort of pre-programming by us (humans). It remains to be seen.

But, researchers and scientists all over the world are working daily to make general artificial intelligence a reality.

You've probably played a lot of different types of card games, including perhaps Go Fish, Gin, Hearts, Spades, Poker (in all its variations), Solitaire (in all its variations), and so on. As you learned these games, you exhibited **general intelligence**. That is, you were able to take your knowledge of a deck of cards – 4 suits, 13 cards in each suit, the hierarchy of the cards in a suit from ace to king, and even some strategies like being void or long in a particular suit – and apply that knowledge to a new card game with a new set of rules. So, you were able to learn the new card game quickly because you already had some experience playing other card games.

It is this type of general intelligence that gives us the ability to quickly learn to perform new intellectual tasks because we can adapt our current knowledge to a new intellectual task. You learned math this way. You first learned to add. You learned to subtract by understanding that, because $3 + 2 = 5$, then $5 - 2 = 3$ and $5 - 3 = 2$. You also learned multiplication as repetitive addition. That is, $8 * 5$ is the same as $8 + 8 + 8 + 8 + 8$. Division was "easy" (relative term) to learn because you already knew subtraction and multiplication.

AGI requires that an AI be able to take its knowledge from one domain and apply and adapt it to another. It requires that an AI adapt to constantly changing environments and completely new and unforeseen circumstances. It requires that an AI foresee a future based on its knowledge of the present and the past.

In short, AGI is human intellect, all of it.

Again, long way off. Probably not in my lifetime, but maybe yours.

From Tic-Tac-Toe to Global Thermal Nuclear Warfare

In the 1983 movie *Wargames*, an AI named Joshua took over the U.S. missile systems with the intent of launching a nuclear attack on the Soviet Union. It simulated many scenarios for doing so and could never come up with a great solution. So, it played tic-tac-toe against itself millions of times and came to understand that the cat always wins. From that, it extrapolated that there could never be a winner in global thermal nuclear war. That's general intelligence to the extreme. (I don't know about you, but once I learned to play tic-tac-toe in such a way that there would never be a winner, I didn't also come to the conclusion that global thermal nuclear war was also a no-win scenario.)

Theory of Mind Artificial Intelligence

Theory of mind artificial intelligence is an AI that interacts socially and can discern and respond appropriately to people's emotions, beliefs, thoughts, expectations, and facial expressions. It takes AI beyond intellectual tasks and into the realm of "understanding" people.

Like artificial general intelligence, we're a long way from theory of mind AI becoming a reality. But, there are a couple of early limited successes in this space. Check out:

- Sophia, a humanoid robot designed by Hanson Robotics. In 2017, Sophia became the first robot to receive country citizenship, that of Saudi Arabia. [3]
- Kismet, a robot head designed by Cynthia Breazeal at MIT, that recognizes facial emotional signals of people. [4]

Sophia and Kismet have many, many videos, articles, and the like. Just do a quick search.

Self-Aware Artificial Intelligence

Self-aware artificial intelligence is an AI that is for all practical purposes a person. It will have emotions and be aware of its emotions. It will be sentient. It will have a conscience. It will understand the need to procreate. It will struggle with ethical decisions. It will probably even act irrationally at times, just as we do, and later apologize.

Self-aware AI is purely hypothetical. We can't even build a self-aware AI with our current set of technologies.

Long, long way off, if we ever get there. And that's a big IF.

Artificial Superintelligence

Artificial superintelligence (ASI) is the most far-reaching view of the potential of AI. An ASI would be the most intelligent entity on the earth, making better and faster decisions than humans in all aspects of life. ASI is popularly referred to as ***singularity***, the point at which machines become smarter than us.

Long, long way off, if we ever get there. And that's another big IF, the same big IF for self-aware artificial intelligence.

The Overlap of the 7 Categories

When I started our discussion of the categories of AI, I did state that the categories were not mutually exclusive. And, indeed, they are not. Think in this way.
- Artificial Narrow Intelligence – a category that includes the (sub) categories of reactive AI and limited-memory AI.
- Artificial General Intelligence – a category of its own.
- Theory of Mind Artificial Intelligence – a category of its own.
- Self-Aware Artificial Intelligence and Artificial Superintelligence – 2 AI category names describing the same type of AI.

I told you... a bit confusing on the various categories.

What's most important to remember is that:
- Artificial general intelligence, self-aware artificial intelligence, and artificial superintelligence are theoretically possible but not yet a reality.
- Theory of mind artificial intelligence has some limited successes with the likes of Sophia and Kismet.
- Artificial narrow intelligence includes almost every successful application of AI. Artificial narrow intelligence can be either reactive or limited-memory.

So, let's spend the rest of our time in artificial narrow intelligence, specifically limited-memory AI.

▲ ▲ ▲

AI BY THE NUMBERS

In 2020, the United States accounted for 40% of the total revenue worldwide in artificial intelligence. [5]

Neural Networks

There are several different types of software tools for implementing AI. We previously talked about an expert system, a tool for creating if-then rules. The expert system follows the rules to produce an outcome. In this way, an expert system is a form of a reactive AI. It cannot adapt itself, create new rules, or learn. It simply *reacts* to the rules and the inputs you provide.

The most common tool for creating limited-memory AI is an artificial neural network. An ***artificial neural network (ANN)*** is a software tool with workings patterned after the human brain, including things like axons, dendrites, neurons, and synaptic connections. (All those wonderful things you learned in your biology class.) While artificial neural network is the correct technical term, you'll often just see the term *neural network*.

Neural Network Layers – Input, Hidden, Output

Let's consider a simple example, that of your wanting to build an AI that can determine if a photo is that of a dog or a cat. (A cat for sure, not guacamole.) You can do so using a neural network like something in Figure 3.2.

Figure 3.2 **Simple Neural Network**

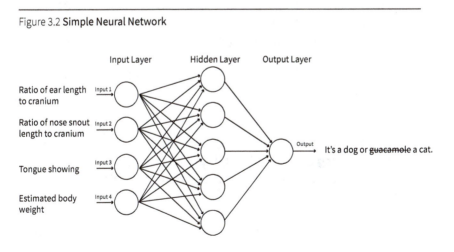

Notice that a neural network has 3 layers:
1. Input – the data you provide to the neural network
2. Hidden – an interior set of nodes or neurons that accept the data from the input layer, mathematically manipulate them, and send a value to the output layer
3. Output – the layer that gives you an answer (i.e., either this is a dog or this is ~~guacamole~~ a cat)

So, you build some image processing software that can look at a photo and gather the following inputs:
- Ratio of ear length to cranium (dogs usually have longer ears in relation to their head size)
- Ratio of nose snout length to cranium (dogs usually have longer nose snouts in relation to their head size)
- Tongue showing (dogs show their tongues more than cats do)
- Estimated body weight (dogs are usually bigger than cats)

Each input node accepts its specific data and passes it on to the interior or hidden nodes. In the hidden nodes is a formula that basically takes the input values, multiplies them by some weights, adds all the results together, and passes the result to the output node.

The output node then sums all the values it receives from the hidden nodes. If the summed value is greater than a threshold value, then the neural network will tell you that the photo is that of a dog. If the summed value is less than a threshold value, then the neural network will tell you that the photo is that of a cat.

Presto. You have a neural network that can distinguish between a cat and a dog. Of course, easier said than done.

Machine Learning

Machine learning describes the ability of an AI to learn from data by changing the way it works. Machine learning gives AI the ability to get better over time, making better decisions based on how it just performed. Using machine learning, you feed your neural network hundreds and hundreds (perhaps even thousands) of photos of both dogs and cats. After you feed in a photo and your neural network "takes a guess," you tell the neural network whether it was right or wrong. This is *training*.

If wrong, the neural network goes back into the hidden layer of nodes and ever so slightly changes one or more of the weights in those nodes to get a slightly different overall value in the output node such that the output node overall value is closer to being on the other side of the threshold value.

If right, the neural network goes back into the hidden layer of nodes and ever so slightly changes one or more of the weights in those nodes to get a slightly different overall value in the output node such that the output node overall value is even farther away from the threshold value (but on the same side of the threshold value).

Either way, this is the concept of *reinforced learning*.

As you can see, the neural network is learning. With every new photo and either a right or wrong answer, it is changing the weights in the hidden nodes to get closer to the right answer (if it guessed wrong) or positively reinforce a right answer. So, the neural network is literally changing how it works by adjusting the weights.

Deep Learning

If you actually implemented our dog versus cat AI, the AI would get pretty good over time, getting better at distinguishing between a dog and a cat with each new photo. It may get to a point that it can perform the task correctly 80 or 90% of the time. To get even more accurate, our neural network would need deep learning.

Deep learning is a subset of machine learning and the ability of an AI to get really, really, really good at making a decision by increasing the number of hidden layers, as in Figure 3.3. (All deep learning is machine learning, but not all machine learning is deep learning.)

Figure 3.3 **Multiple Hidden Layer Neural Network**

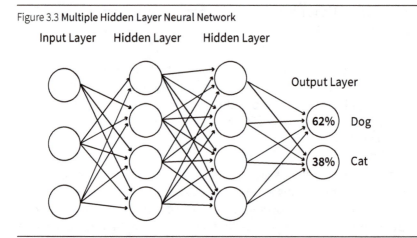

(Opting to break out the weed whacker at this point.) The more hidden layers you have, the more the neural network can take into account and work with the subtleties and nuances within the data. It's rather like adding another layer to a decision-making process.

So, deep learning can help create a more effective AI. But, deep learning does require much more learning data and a longer learning process. If you double the number of nodes in the hidden layer, it makes sense that more training will be required to get the weights set appropriately.

As long as we're talking about dogs, a great example of the need for deep learning would be to build an AI that can distinguish between all the different breeds of dogs. Do you know how many there are? Well, depending on who you ask, there are somewhere between 190 and 360. (I know, it seems odd that we can't even come up with an exact number.) To distinguish between all of these, we would need a deep learning AI to work with all the subtleties of the very slight differences among so many breeds.

Supervised and Unsupervised Learning

In our dog/cat example, we used supervised learning. **Supervised learning**, in artificial intelligence, is a training process in which you specify the inputs to be used and then also identify whether an outcome is right or wrong. Again, in our example, we constrained the AI to just using the 4 inputs we specified. But, there are most likely other characteristics that may help distinguish between cats and dogs... paw size, sharpness of nails, something about the tail, perhaps something about fur... you get the idea.

So, we could have opted for **unsupervised learning**, a training process in artificial intelligence in which you do not specify the inputs and you simply let the AI determine the best sets of inputs to use to arrive at a correct answer.

For example, our dog/cat AI may start to take into account the setting of the photo, either indoor or outdoor. I'm just guessing here, but if you took all the photos of cats and dogs on the Internet, I bet the biggest proportion of dog photos leans more toward an outdoor setting while the biggest proportion of cat photos leans more toward an indoor setting.

Collar presence may be another. If you take all the photos of cats and dogs on the Internet, what do you think? More dog photos with collars or more cat photos with collars? Interesting.

It's really difficult to explain unsupervised learning with just words on a page. So, let me make a recommendation for a great illustration. Visit https://www.youtube.com/watch?v=qv6UVOQ0F44&t=128s (YouTube video) and watch how an unsupervised neural network learned to master Super Mario World in just 24 hours. Truly, truly fascinating to see how a neural network mastered the game with only the simple instruction of maximizing fitness.

▲ ▲ ▲

Learning from Mother Nature

To facilitate the learning process – either supervised or unsupervised – we've **learned** from mother nature, the most sophisticated natural learning intelligence to ever exist. Our "mother nature" concepts include genetic algorithms, neuroevolution, and biomimicry.

Genetic algorithms are a set of techniques modeled after biologically-inspired intelligences that primarily include the following:
- Selection – choosing a better outcome over a poorer outcome
- Crossover – combining 2 "good" outcomes to see if a better outcome can be achieved
- Mutation – randomly trying something new to determine if a better outcome can be achieved

The thinking is quite simple. In selection, a better outcome is chosen over a less optimal outcome. In crossover, an attempt is made to combine two good outcomes to create a better outcome. In mutation, just try something "new" to see if you get better results. You've probably done each of these while playing a video game. You've opted for one strategy over another because the first yields better results (selection). You've combined the simultaneous pressing of 2 buttons to see if you get a better result (crossover). And you've probably gotten stuck and just randomly tried something new to see if you can get "unstuck" (mutation).

These are all concepts we can implement in artificial intelligence.

Neuroevolution is the use of genetic algorithms within an unsupervised learning context. With neuroevolution, you allow things like selection, crossover, and mutation to be the primary drivers of how a neural network learns without specifying the exact inputs, or potentially even the outputs to a given set of inputs. Hard to say more without needing the old weed whacker… check out the Super Mario example on YouTube I mentioned earlier.

Biomimicry is the simulation and use of elements of nature to solve human problems. For example, we've learned much from hiving insects (bees, wasps, etc.) to determine how to better build things like buildings and homes. We've studied forager ants to determine how to better design supply chain management activities. We studied birds to build better the aerodynamic nature of airplanes. We studied termite mounds to better understand air flow and ventilation.

Fascinating field. You should do some further research into the study of biomimicry and its applications.

▲ ▲ ▲

Building Your Own AI

You can build your own AI, although I will tell you that artificial intelligence development is among some of the most difficult to do. You'll most likely need to a learn a specialized programming tool that supports AI development. Below are some recommendations. [6]

- Scikit-Learn – https://scikit-learn.org/stable/
- TensorFlow – https://www.tensorflow.org/
- Caffe – https://caffe.berkeleyvision.org/
- Theano – https://pypi.org/project/Theano/
- Keras – https://keras.io/
- MXNET – https://mxnet.apache.org/versions/1.8.0/
- Microsoft CNTK – https://docs.microsoft.com/en-us/cognitive-toolkit/
- Google AutoML – https://cloud.google.com/automl
- Accord.NET – http://accord-framework.net/
- Torch/PyTorch – https://pytorch.org/

If you review any of these, you'll notice that many are Python-based. So, I would definitely recommend that you have some Python expertise.

▲ ▲ ▲

Biases in Artificial Intelligence

Unfortunately, you can build biases into an AI without even realizing it.

Think about building an AI to determine something like credit approval, amount of a loan, or some other financially-based decision. You thoughtfully decide to eliminate inputs around demographics... age, gender, race, and ethnicity. And instead, you choose to include education level, length of employment, salary, and amount of savings.

Your thinking is simple. A better credit risk is someone with a higher level of education, a longer length of employment, a higher salary, and a higher amount of savings. That does make sense. But <u>who</u> have you just described? To be blunt and completely honest, you've described the typical white middle-aged male. Without explicitly programming demographic inputs, you have *implicitly* described a specific group of people, a group of people who have benefited from their demographic status and, as a result, typically have a high level of education, more consistent employment, a comparatively high salary, and a comparatively higher amount of savings.

Please understand everyone that I am in that group of people who have benefited from their demographic status. I'm not casting aspersions. I'm simply stating fact. Even if you eliminate demographic inputs, you must understand that demographic inputs often have a direct correlation to other (often behavioral) inputs.

This is a hugely important issue we must address in the use of artificial intelligence.

▲▲▲

Applications of Artificial Intelligence

Okay, there are way too many of these to list here. AI, specifically as artificial narrow intelligence, has been around for some time and is very successful. Let me cover a few of those here for you. [7, 8]

- Roomba, by iRobot – Roomba uses AI to scan the room for furniture and other types of obstacles, determine the room size, and build a memory of the most efficient path for cleaning a room.
- Olly, by Emotech – Olly is an AI assistant like Alexa but has an evolving personality.
- Covera Health – uses AI to reduce the number of misdiagnosed patients. The AI sifts through previous patient histories to provide a better profile of symptom information.
- Autonomous vehicles – duh. (Read Chapter 6.)
- Betterment – an AI financial advising platform. It uses AI coupled with the profile of an investor and a wide array of financial investment alternatives to recommend an investment portfolio tailored to the investor.
- Hopper – uses AI to help predict when you, as a consumer, can get the lowest prices on flights, hotels, and other travel-related expenses.
- Social media – from Meta to Twitter to Slack, all social media platforms are running AI to better match people and content.
- Grammarly – uses AI to improve your writing. (Again, duh.)
- Smart thermostats – use AI to learn occupants' behaviors and adjust the thermostat accordingly.
- KenSci – has built an AI platform to help identify fraudulent insurance claims.
- Biometrics – many applications of AI in biometrics including facial recognition.
- Supply chain management – all the big delivery players – Amazon, UPS, USPS, FedEx, DHL, etc. – are using AI to determine the optimal routes for delivering packages, how to optimize the loading of trucks, and many other aspects of delivery.

▲ ▲ ▲

The Web 3.0

The Web 3.0 isn't specific to or a subset of artificial intelligence. Which begs the question, "Why, then, is it in the chapter on AI?" (Another great question on your part.)

In formulaic terms, the Web 3.0 looks like this.

Web 3.0 = Web + AI + DLT + IoT
where DLT includes:
Blockchain
Cryptocurrency
Tokenization
Smart Contracts
Dapps
DeFI

Now that's a weed-whacker mouthful, for sure, so let's try to break it down.

The Evolving Web

Web 1.0, the first iteration of the Web, was characterized by few creators of content (usually large organizations) and many users or consumers of content. Web sites and pages were mostly static (meaning their content often had to updated manually by changing the underlying HTML code). We're well beyond Web 1.0.

Web 2.0, the current iteration of the Web, is characterized by massive amounts of user-generated content (blogs, reviews, social media, etc.), with all content being dynamically delivered and updated in real-time, often responding to actions and requests initiated by users. Web 2.0 is often called the *participative social web*. But "social" is just one aspect of the dynamic nature of the Web 2.0. We can watch sports updates in real time, changing bids on eBay, etc. Basically, we can see content on the Web today as that content emerges, changes, and eventually dies off.

The **Web 3.0** is the next evolution beyond Web 2.0 that will exhibit semantic characteristics (AI), decentralized protocols (distributed ledger technologies), and be ubiquitous (Internet of Things). Of course, now we need to chat about semantic, decentralized protocols, and ubiquity. (I see the weeds growing.)

Semantic

Semantic refers to a concept called metadata, essentially data about data. Using metadata, Web 3.0 applications will be able to – in automated fashion – connect related data sources from all over the world. This will require the use of artificial intelligence to make sense of and interpret the data. Consider these:

- I love Bitcoin
- I <3 Bitcoin
- I heart Bitcoin
- I ♥ Bitcoin

Syntax and use of words and symbols are different; the semantic meaning of them is essentially the same. We can understand that; computers will need AI to do so.

Decentralized Protocols

As we discussed in Chapter 2, most information storage and processing is still centralized. The Web 3.0 requires the decentralization of information and software using distributed ledger technologies like blockchain, cryptocurrency, tokens such as NFTs, smart contracts and Dapps that automate tasks, and decentralized financial (DeFi) applications.

Ubiquity

Ubiquity simply means everywhere. For the Web to be everywhere, everything must be connected to the Web. Which means that the Internet of All Things will play an important part in the foundation of the Web 3.0. Connecting all things – computer-related, electronic, and non-electronic – to the Web will create a Web 3.0 that is everywhere.

We're definitely not at Web 3.0 yet. But we'll eventually get there in the 4th industrial revolution. As we evolve to Web 3.0, you probably won't even notice the changes as they occur. They will be subtle and small. But, over the next 10 years, the Web will change dramatically, with all those subtle and small changes accumulating into a significant shift in the Web.

▲ ▲ ▲

IMAGINE FORWARD

In the Next 10 Years, You'll Work Along Side an AI

I firmly believe that most all jobs in the next 10 years will involve working with an AI. And I'm not talking about just "professional" jobs like financial managers, product planners, and marketing mix managers. I'm also talking about people who work the production floor in a manufacturing environment… all of us (more appropriately, all of you) will have an AI as a colleague, in some form or fashion.

You'll Need a Course in AI Resource Management

Right now, most every business program in the country includes a course in human resource management, how to effectively manage people to maximize their productivity, enable personal and professional development, resolve conflicts, etc. Future business programs will include a course in AI resource management.

An AI may very well come to act like people. It may have a bad day. It may not want to come to work. It may be distracted by world events.

I know those statements seem way far-fetched. But I'm not so sure. Theory of mind AI progress is at a feverish pace.

Add Just a Pinch of AI

As we discussed in Chapter 1 and the Internet of All Things, we'll see AI in most all settings. Bathtubs that stop filling when the water level gets too high… a simple reactive AI. Books that automagically generate more problems for you to work because you're struggling with a particular topic.

To start innovating in this space, go back to "add just a pinch of…" The pinch in this instance is AI, obviously. To what existing products and services can you add some AI? What would be the benefit? How would the AI get trained?

Job Loss & Gain

Throughout the 275 years of industrialization, there have always been concerns that new technological advances will create unemployment. Partially true to be sure. An IBM survey in 2019 estimated that 120 million workers globally would need retraining in the following 3 years because of AI's impact on jobs. [9]

But each new technological invention and revolution has created more jobs (and increased our quality of life). No doubt, many jobs will be lost to AI, but even more will be created. We've been through this many times before. The telephone and telegraph ended the employment of the riders for the Pony Express. Advances in communications technologies and software displaced most telephone operators. But, in each instance, more new jobs were created.

Our willingness to invest in retraining is a must.

TOP COMPANIES TO WATCH IN AI

- CCC
- Veritone
- Apple
- LogicMonitor
- Google
- nate
- People.ai
- Ascent
- Osaro
- Riskified
- Nvidia
- Amazon
- Nuro
- Tempus
- Tesla
- Ascent
- DataRobot

- veda
- Urbint
- Circulo Health
- Moveworks
- AlphaSense
- Narrative Science
- Clarifai
- Neurala
- nuTonomy
- Persado
- CognitiveScale
- AEye
- AIBrain
- Blue River Technology
- Freenome
- Grammarly
- CloudMinds

- Vidado
- Casetext
- CloudMinds Technology
- Insilico Medicine
- Nauto
- Verint Next IT
- Orbital Insight
- Sherpa.ai
- SoundHound
- Vicarious
- Zebra Medical Vision
- Zoox
- Zymergen
- H20.ai
- Nauto
- OpenAI
- Sift Science

Some are big and well-known, some are chasing AI across a broad spectrum of applications, some are emerging, and some are specific to an industry. [11]

AI BY THE NUMBERS

According to a survey of 1,032 businesses, 67% believe AI can create better customer experiences, and 54% believe AI can improve decision making.
[10]

Closing The Divides

Big Hairy Audacious Problems

Opportunities for innovation within the 4th industrial revolution are enormous and most probably immeasurable. New product and service offerings will be fascinating to see.

But, your focus shouldn't be solely profit-motivated.

Indeed, there are many problems in the world that deserve your attention and that can be solved using innovations in the 4th industrial revolution.

We all have a moral and societal obligation to address these problems. I've listed a few below with some startling statistics.

The Education Divide

- 72 million – number of children around the world who remain unschooled. [1]
- 32 million – number of primary age school children uneducated in Sub-Saharan Africa. [2]
- 54% - girls as a percent of the non-schooled population in the world. [3]
- 80% - percent of girls in Yemen who will never have the opportunity to go to school. [4]
- 258 million – children and youth who are out of school worldwide. [5]
- The wealthiest 10% of U.S. school districts spend nearly 10 times more per student than the poorest 10%, and spending ratios of 3 to 1 are common within states. [6]
- In predominantly minority schools in the U.S., which most students of color attend, schools are large (on average, more than twice as large as predominantly white schools and reaching 3,000 students or more in most cities); on average, class sizes are 15% larger overall (80% larger for non-special education classes); curriculum offerings and materials are lower in quality; and teachers are much less qualified in terms of levels of education, certification, and training in the fields they teach. [7]
- 50% - chance of getting a math or science teacher with a license and a degree in the field in U.S. schools with the highest minority enrollments. [8]

The Food Divide and Food Waste

- 35% - percent of the 229 million tons of food available that went unsold or uneaten in the U.S. in 2019. [9]
- 38 million – number children in 2019 who were overweight or obese. [10]
- 25,000 – number of people who die daily from hunger and related causes. [11]
- 10,000 – number of children who die daily from hunger and related causes. [12]
- 854 million – number of people worldwide who are undernourished. [13]
- 29.6 million – number of children in the U.S. each day who receive low-cost or free lunches. [14]
- 21.3% - rate of stunting (children too short for their age as a result of chronic malnutrition) among children in the world in 2019. [15]

The Health Care Divide

- 50% - percent of people worldwide who cannot obtain essential health services. [16]
- 800 million – number of people who spend at least 10% of their household budgets on health care expenses. [17]
- 17% - percent of mothers and children in the poorest fifth of households in low- and lower-middle income countries who receive at least 6 basic maternal and child health interventions. [18]
- Children born in Sierra Leoni, West Africa, have a life expectancy of 50 years, compared to Japan where the life expectancy is 84 years. [19]
- In 2019 in the U.S., 21.7% of AIANs (American Indian and Alaska Native), 20% of Hispanics, and 11.4% of Blacks had no health insurance. Among Whites, the figure was 7.8%. [20]
- In the U.S., Black adults are most likely to have obesity (47.5%), followed by Hispanic adults (46.9%). White adults are at a rate of 38.2% [21]
- 1 in every 10 LGBTQ+ people reported that a health care professional refused to see them because of actual or perceived sexual orientation. [22]

The Wealth Divide

- The world's 2,153 billionaires collectively have more wealth than the world's 4.6 billion poorest people. 4.6 billion is 60% of the world's population. [23]
- The 22 richest men in the world have more wealth than all the women in Africa. [24]
- Black families in the U.S. own 3% of total household wealth, despite making up 15% of households. White families own 85% of total household wealth, despite making up only 66% of households. [25]
- In 2020, the average CEO compensation at the largest low-wage corporations was almost $14 million annually. The median hourly annual wages at those corporations was a little over $30,000. [26]
- The richest 0.1% of Americans earn 196 times as much as the bottom 90% of Americans. [27]
- The top 10 countries by GDP per capita range from $180,000 (Liechtenstein) to $63,000 (United States). There are 26 countries with a GDP per capita lower than $1,000. [28]
- 9.2% - percent of the world's population – roughly 700 million people – who survive on less than $1.90 per day. [29]
- Countries with the highest poverty rates include South Sudan (82%), Equatorial Guinea (77%) and Madagascar (71%). The U.S. poverty rate is 13.4%. [30]

The Utilities Divide

- 2.2 billion – number of people worldwide who do not have safely managed drinking water services. (3 billion lack basic handwashing facilities.) [31]
- 15% - percent of the world's population who did not have access to electricity in 2016. [32]
- Indonesia is close to total electrification (98%), while Chad's is only 8.8%.[33]
- At the onset of the COVID pandemic, 3 of every 10 people worldwide could not wash their hands with soap and water in the their homes. [34]
- In 2020, nearly half the world's population lacked safely managed sanitation. [35]
- Worldwide, women and young girls spend 266 million hours annually trying to find a place to get water. [36]
- Nearly 1 million – number of people who die each year from water, sanitation, and hygiene-related diseases. [37]

And, of course, those are just a very few of the many startling statistics regarding education, food, health care, wealth, and utilities.

There many other significant inequities, divides, and problems that your generation needs desperately to address, very quickly. Some of those include the likes of human trafficking, ocean pollution, and global warming.

It is my hope that you will take time to consider innovating in the 4th industrial revolution to address the world's most important issues.

United Nations Sustainable Development Goals

The United Nations is currently working to address the following 17 sustainable development goals. [38]

1. No Poverty – End poverty in all its forms everywhere.
2. Zero Hunger – End hunger, achieve food security and improved nutrition and promote sustainable agriculture.
3. Good Health and Well-Being – Ensure healthy lives and promote well-being for all at all ages.
4. Quality Education – Ensure inclusive and equitable quality education and promote life-long learning opportunities for all.
5. Gender Equality – Achieve gender equality and empower all women and girls.
6. Clean Water and Sanitation – Ensure availability and sustainable management of water and sanitation for all.
7. Affordable and Clean Energy – Ensure access to affordable, reliable, sustainable, and modern energy for all.
8. Decent Work and Economic Growth – Promote sustained, inclusive, and sustainable economic growth, full and productive employment and decent work for all.
9. Industry, Innovation and Infrastructure – Build resilient infrastructure, promote inclusive and sustainable industrialization and foster innovation.
10. Reduced Inequalities – Reduce inequality within and among countries.
11. Sustainable Cities and Communities – Make cities and human settlements inclusive, safe, resilient and sustainable.
12. Responsible Consumption and Production – Ensure sustainable consumption and production patterns.
13. Climate Action – Take urgent action to combat climate change and its impacts.

14. Life Below Water – Conserve and sustainably use the oceans, seas and marine resources for sustainable development.
15. Life on Land – Protect, restore and promote sustainable use of terrestrial ecosystems, sustainably manage forests, combat desertification, and halt and reverse land degradation and halt biodiversity loss.
16. Peace, Justice and Strong Institutions – Promote peaceful and inclusive societies for sustainable development, provide access to justice for all and build effective, accountable and inclusive institutions at all levels.
17. Partnerships for the Goals – Strengthen the means of implementation and revitalize the global partnership for sustainable development.

Within those 17 goals, the U.N. has established a total of 169 measurable targets.

CHAPTER 4

Extended Reality

No Matter Where You Don't Go, There You Are

Extended Reality

Extended reality is an umbrella term for augmented reality, mixed reality, and virtual reality, which are all some combination of the physical and virtual worlds through technology.

Consider a continuum from the completely physical to the completely virtual (See Figure 4.1). On the far left side is the completely physical (real) world, with no technology enhancement. That's basically experiencing life without the aid of technology. (If you haven't recently, you should try it. The purely physical world is wonderful.) On the far right side is a completely virtual world, what is commonly called *virtual reality*, something you've most likely already experienced.

Figure 4.1 **The Physical and Virtual Continuum**

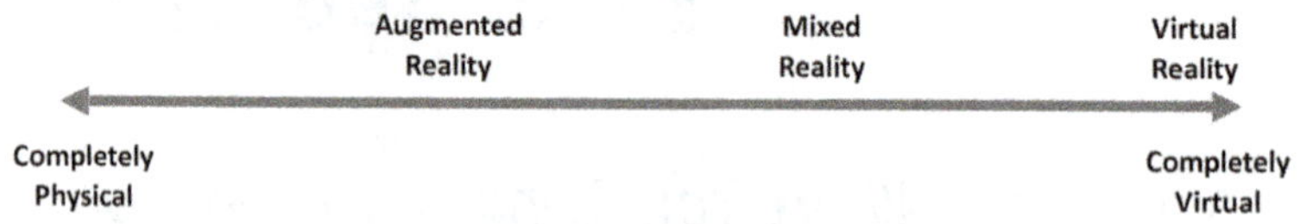

In between are augmented and mixed reality. Augmented reality is based mostly on the physical world with technology-enhanced content and information. Mixed reality is the combining of the physical and virtual worlds in which the two must interact, be constrained by each other, and "fit" together.

▲ ▲ ▲

Augmented Reality

Augmented reality (AR) is the enhancing of your view of the physical world by adding content, information, and the like.

Augmented reality is already very much a part of your life, for example, while watching sports on TV. If you're watching Olympic swimming, you'll see added lines that mark the world record and Olympic record time splits. If you're watching (American) football, you'll see lines that show the distance to a first down or where the team needs to get for a field goal. Replays of soccer goals will show ball trajectory and speed. The list goes on and on.

So, augmented reality is about taking an image (video or photo) of the physical world and adding content to it to enhance or **extend** what you're seeing.

Your generation played Pokemon Go, which is augmented reality-based. Go the appropriate location and use your phone to see and capture characters. Many restaurants now have AR-enhanced menus. Use your phone to view a menu item and content will pop up showing the ingredients, how it's cooked, and so on. Google Translate can translate (duh) content, text, and road signs as you view them through your camera.

So, augmented reality exists mostly in the physical world, with added (virtual) content to enhance what you are seeing.

What Technologies You Need

You don't really need any "special" technology to take advantage of almost all augmented reality applications, just your camera or a viewing screen like your TV.

Not much more to say.

Building Augmented Reality Applications

If you're interested in learning how to build augmented reality applications, consider these.

- Unity (www.unity.com) – Unity was developed as a tool for game development, first for iOS and then extended to Android. It was the tool used to create the likes of Pokemon Go and Call of Duty: Mobile. It's become a very popular tool for developing games, augmented reality applications, and virtual reality applications. It's also a great tool if you're just learning to build augmented reality applications, and it's very robust for advanced developers.
- Vuforia (www.vuforia.com) – Vuforia is a cross-platform augmented reality development tool, meaning that you can build for both iOS and Android devices within a single tool. It's integrated with Unity. If you search for courses in Udemy for AR development, you'll find many that teach a combination of Unity and Vuforia.
- ARKit – ARKit is Apple's AR development tool for iOS.
- ARCore – ARCore is Google's AR development tool for Android.

Obvioulsy, ARKit and ARCore are specific to a technology platform, either iOS or Android. Unity and Vuforia support cross-platform development, so they get my recommendation. You can learn the basics of both in Udemy for about $100 (sale prices can go down to around $20 for a course) in less than 20 hours. Not a bad way to spend a few days.

What are you going to put on your resume… that you know Microsoft Office tools like Word and Excel, or that you can build augmented reality applications?

Applications of Augmented Reality

There are many great examples out there of augmented reality.

In the app world, take a look at Google Maps, Civilisations AR, Froggipedia, The Machines, Smash Tanks, Big Bang AR, Vuforia Chalk, Thyng, Insight Heart, Houzz, and IKEA Place. [1] The latter 2 let you pick furniture and see how they fit into your home, apartment, or dorm room.

Enterprise AR is going to be big business. *Enterprise AR* focuses on the use of AR for organizations. [2]

- Remote Assistance – helping users troubleshoot problems and perform maintenance and repair (Mercedes-Benz, Vodaphone, and Prince Castle).
- Medical Diagnostics – diagnosing diseases and illnesses (Aris MD, OrcaHealth, and surgeons in Brazil).
- Marketing and Sales – enhancing marketing material, moving through the sales process, providing post-sales support (Go-Digital, Cannondale, and Andersen Corporation).
- Training – not much explanation needed here (AGCO, Japan Airlines, and Siemens).
- Logistics – improving daily operations in warehousing, transportation, inventory management, etc. (Amazon, BMW, and DHL).
- Manufacturing – identifying and resolving issues on the manufacturing floor (VitalEnterprises, Airbus, and Bosch).
- Prototype Design – building and reviewing prototypes of new product designs (Aecom, AirMeasure, and BNBuilders).

▲ ▲ ▲

EXTENDED REALITY BY THE NUMBERS

The global virtual reality market is expected to be $70 billion in 2028, up from $16 billion in 2020. [3]

Virtual Reality

On the other end of the spectrum is virtual reality. ***Virtual reality (VR)*** is a completely immersive simulated virtual experience, which may be based in a real world environment or a fictitious one. So, you can use virtual reality to experience walking the Great Wall of China or swimming the Great Barrier Reef off the coast of Queensland, Australia. You can also experience the wreckage of Jurassic Park (Jurassic Park: Afterworld) or you can leap, slide, run, and jump across city rooftops (Stride). The Great Wall of China and the Great Barrier Reef are based on real world environments. Jurassic Park and Stride's city rooftops are based on fictitious ones.

The focus of virtual reality is to transport you via your senses to another place and set of experiences. Your senses here include what you see, what you hear, and what you feel. (Feeling also works in reverse, as the VR system will detect your movements and respond appropriately.) And, some VR applications are starting to incorporate the sense of smell. Check out OVR Technology at www.overtechnology.com.

What Technologies You Need

Most people your age have experienced virtual reality, as it's a very popular platform for gaming, including Arizona Sunshine, Batman Arkham VR, Astro Robot: Rescue Mission, Beat Saber, Elite Dangerous VR, Fallout 4 VR, Half-Life: Alyx, Lone Echo 2, Minecraft VR, Pistol Whip, Resident Evil 4 VR, and many, many, many, others.

To experience VR, at a minimum you'll need a VR headset (an electronic headset that adjusts its field of vision to your moving your head in different directions, up and down and side to side) and some sort of hand controller for selecting menu items, pointing at things, and other interface control tasks. While hand controllers can perform very basic functions like those I just listed, they can also be used as specialized controllers depending on the situation, for example, a fishing pole, a weapon, a bowling ball, a light saber, etc.

Popular VR headsets include:
- HTC Vive Cosmos Elite
- Oculus Quest 2
- Sony Playstation VR
- HTC Vive Pro Eye
- HP Reverb G2
- Valve Index VR Kit
- Oculus Rift S

Of course, these are changing and being updated all the time. By the time you read this, for example, Oculus Quest 2 may be Oculus Quest 3.

If you haven't yet, I would recommend you try out Google Cardboard. It costs only $10 and isn't electronic at all. Instead, you drop your phone into the Cardboard and your phone becomes your screen. A very inexpensive way to experience virtual reality.

Building Virtual Reality Applications

You can build your own VR applications but it's definitely more complicated than building augmented reality applications.

Many of the VR development tools can also help you build augmented reality applications. If you're interested in building VR, I would recommend that you start by building some AR applications first, because they're easier. As you get more comfortable with the tool, you can then move on to VR development. VR development tools include: [4]
- Unity
- Amazon Sumerian
- Google VR for Everyone
- Unreal Engine 4
- CRYENGINE
- Blender
- 3ds Max
- SketchUp Studio
- Maya
- Oculus Medium

Several of these also pop up in the 3D printing space, which we'll discuss in the next chapter.

Applications of Virtual Reality

As with augmented reality, applications of virtual reality abound.

In the personal space, I already mentioned games including Arizona Sunshine, Batman Arkham VR, Astro Robot: Rescue Mission, Beat Saber, and many others. Consider also checking out Google Expeditions, Colosse, Titans of Space, Google Earth VR, Kingspray Graffiti VR, Ocean Rift, Foo Show, Virtual Desktop (your computer in VR), and Within.

On the enterprise front, all organizations seem to be racing to take advantage of virtual reality. [5] For each area below, I've included one or a few examples, but there are many others.

- Online Shopping and Retail (IKEA Reality Kitchen Experience for designing your perfect kitchen)
- Engineering and Manufacturing (Airbus for determining the comfort of seats)
- Tourism and Hospitality (virtually experiencing museums, resorts, and travel options)
- Communication and Collaboration (virtual team meetings)
- Training (sales, military, and many more)
- Architecture, Construction, and Design (what the interior of a building will look like before building it)
- Medical and Health Care (treating psychological disorders, practicing surgery on virtual cadavers)
- Education (experiencing as opposed to listening to a lecture)
- Sports (athletes experiencing simulated training sessions and in-game situations)

▲ ▲ ▲

Mixed Reality

In between augmented and virtual reality is mixed reality. **Mixed reality (MR)** is the merging of the physical and virtual worlds in which both worlds simultaneously co-exist in real-time and are constrained by each other.

Let's consider a simple example, that of designing your dorm room, focusing on the optimal layout of your furniture. Using mixed reality, you would first use your phone to *map* the room. You can now do this with the latest phones that have Lidar capabilities. In mapping your room, you're getting the dimensions of the room, including things like the height of the window, how far the door is from each wall, how wide the door is, the exact location of the sink (if you're lucky enough to have one in your room), and so on.

You would also use your phone in similar fashion to determine the dimensions of the existing furniture. Then, within mixed reality, you could move the furniture around virtually. You could virtually add new furniture, with the mixed reality not allowing you to block the door with a chair, for example.

I think you get the idea. Mixed reality is a combination of the physical (the dimensions of your room) and virtual world (the dimensions of furniture you're thinking about adding) with the physical constraining the virtual and vice-versa.

What Technologies You Need

To experience mixed reality, you can use your phone and some apps like Houzz or IKEA Place to do what I just described in designing your dorm room.

More realistically, you'll want a mixed reality headset. A mixed reality headset is similar to a VR headset, except that a mixed reality headset includes the mapping capability of a room or environment. The dominant mixed reality headsets include: [6]
- Lightfield (Avegant)
- AjnaLens (Dimension NGX)
- One (Magic Leap)
- HoloLens & HoloLens 2 (Microsoft)
- Light (NReal)
- Varjo XR-3 (Varjo)
- Bridge (Occipital)

Building Mixed Reality Applications

In the mixed reality space, you'll need to learn development skills particular to the headset hardware you choose to use. I listed them above.

Microsoft and Magic Leap dominate the mixed reality space, so consider going with either of those if you want to pursue a career in mixed reality development. Starting salaries are in the $150,000 - $200,000 range. Very nice.

By the way, if you do choose to learn mixed reality development, you'll also be learning augmented and virtual reality development. So, consider perhaps learning basic augmented reality development first as a lead in to what can be very complex development in mixed reality.

Applications of Mixed Reality

Mixed reality is the newest of the three in the extended reality space. To get a feel for mixed reality, I would recommend that you watch the following videos:
- Envisioning the Future with Windows Mixed Reality (https://www.youtube.com/watch?v=2MqGrF6JaOM&t=2s)
- HoloLens 2 AR Headset: On Stage Live Demonstration (https://www.youtube.com/watch?v=uIHPPtPBgHk)
- Introducing Microsoft Mesh (https://www.youtube.com/watch?v=Jd2GK0qDtRg)
- You can also find examples in:

- Medical Education
  (https://www.youtube.com/watch?v=h4M6BTYRlKQ)
- Gaming (Star Wars, Game of Thrones, Angry Birds)
- Merchandising (for determining optimal store layout virtually before physically moving everything around, over and over again)
- Urban planning (for understanding the impact of new buildings, bus stops, etc.)

I think one of the big opportunities for mixed reality is in the team collaboration space. We can already have virtual meetings, bringing people together from all over the world. But, with mixed reality, we can actually bring people virtually into the room with avatars and holograms. Using your mixed reality headset, you'll be able to see 3D images or holograms of your co-workers in the room with you.

EXTENDED REALITY BY THE NUMBERS

In 2020, 53% of all virtual reality applications were in the commercial space. Businesses are quickly capitalizing on the benefits of VR in spaces like retail, car showrooms, and real estate.
[7]

Complexity of Development – AR, VR, and MR

In terms of complexity of development, augmented reality is the easiest, mixed reality is the most difficult, and virtual reality is somewhere in between (See Figure 4.2). It makes sense when you think about it.

Augmented reality is about recognizing images and the like through a camera and then laying content on top of those images. Building the very basic version of Pokemon Go isn't that difficult. You need to access the location of the user via their phone GPS, track the user to a specific location, present an image of a character on the phone's screen, and enable the user to capture the character in some way.

Figure 4.2 **Extended Reality (XR) – Complexity of Development**

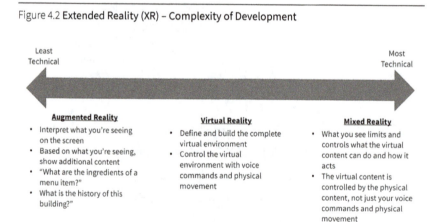

Virtual reality is more challenging because you have to build a complete virtual world and capture and respond to a lot of movement by the user (looking in different directions, hand movements, etc.). You also have to build the virtual environment on a *movement continuum*. That is, as the user progresses through a virtual environment, you have to have built the virtual environment for the next location. Think about a virtual reality in which you move the user through a building. You have to have all rooms of the building built and continually change the view of the building as the user moves through it.

Mixed reality is the most challenging to build because you have to map the physical world around the user and use that mapping to constrain what the user can do in the virtual world. If you're building a mixed reality for someone to virtually place different pieces of furniture in their home, you have to map the size of the room and use those dimensions to ensure that the furniture will fit according to where the user wants to put it.

▲ ▲ ▲

EXTENDED REALITY BY THE NUMBERS

The global augmented reality market is expected to be $340 billion in 2028, up from $18 billion in 2020. [8]

Metaverse – Extended Reality Extended

Metaverse is a virtual universe – aided by technologies such as augmented, virtual, and mixed reality, video, holograms, avatars, and various sensing technologies – in which people connect, live, work, conduct commerce, and play. This can include team collaboration in a business environment, virtual trips, going to conferences and concerts, and even something as simple as having a conversation with an avatar- or hologram-based person anywhere in the world.

Metaverse has gained considerable attention most recently because of Facebook's and Microsoft's big push into the space. In late 2021, Facebook changed its name to Meta, signaling a significant shift into the metaverse space. According to Mark Zuckerberg, "The next platform and medium will be even more immersive and embodied Internet where you're in the experience, not just looking at it, and we call it the metaverse." [9]

During that same year, Facebook/Meta acquired Within, Unit 2 Games, Bigbox VR, and Downpour. All those are in the metaverse space in some form or fashion. (Facebook acquired Oculus in 2014.) [10]

About the same time, Microsoft announced that it was adding HoloLens and its mixed reality capabilities to Microsoft Teams, a collaboration tool that many businesses use. The tool is called Mesh. Mesh allows you to participate in virtual meetings as your actual self or a 2D or 3D avatar or hologram. And you won't need a headset to view other people who choose to go the 3D avatar route. [11]

There a number of other companies already in this space or emerging including:
- Decentraland
- IMVU
- Epic Games
- Roblox
- Minecraft
- Second Life

- Nowhere
- The Sensorium Galaxy
- Super World
- Sandbox

The goal of the metaverse concept is to connect people through a more immersive experience than traditional technology tools like social media and team collaboration tools.

We're going to see entire cities emerge in metaverse. You can already use cryptocurrency to buy NFTs that represent land ownership in metaverse. You can use more cryptocurrency to build on your land, perhaps an amusement park or a concert venue.

You can expect to see "traditional" businesses like car manufacturers setting up shop in metaverse. You'll have an immersive experience test driving a car and determining which features you want. After you've put together your package, you can transfer that to a local dealership and buy the exact car you want.

Even governments are getting into metaverse. In late 2021 the country of Barbados announced it was building an embassy in Decentraland, a popular crypto-powered metaverse. Barbados even declared its embassy land purchase to be sovereign land. Barbados will offer such services as e-visas and the ability to extend cultural diplomacy, the trading of art, music, and culture. [12]

Metaverse is one of the many new "wild, wild wests" in the 4th industrial revolution. There will most certainly be plenty of money made and lost in metaverse. Whether you aspire to be an intrapreneur or entrepreneur, you should consider evaluating metaverse as an investment opportunity.

▲ ▲ ▲

IMAGINE FORWARD

You Won't Need to Build Augmented, Mixed, and Virtual Reality

As I stated, you'll obviously have to learn to build if you want to pursue a career in extended reality development. But, for the typical consumer or user of extended reality, we're going to have apps and embedded extended reality.

As for extended reality apps, this is going to be no different from today's phone apps. You don't have to write your own text messaging app; it's already on your phone. You don't have to write your own video-based social media app; you can download it from the app store.

The same will be true for extended reality.

So, we'll have Vapps (virtual reality apps), Mapps (mixed reality apps), and Aapps (augmented reality apps… I'm not real sure how to pronounce *Aapps*, but fortunately pronunciation isn't important when writing a book).

Phones, tablets, and computers will come with built-in capabilities in the extended reality space. We're already seeing that with the newer phones and tablets that incorporate Lidar and augmented reality enhancement capabilities.

You'll be able to download just about any environment – real or fictitious – for virtual reality. You can almost do that now.

Mixed reality is going to be the really interesting space. I think it becomes a personal productivity tool with built-in libraries or capabilities and images. Think about Microsoft Word (or whatever word processing tool you use). It's a personal productivity tool for creating text-based documents. You can create your resume, an advertising flyer, term paper, and so on. You don't need different word processing software for each situation.

I think the same will be true for extended reality. You'll get libraries of images, objects, enhancements, and effects. Your mixed reality headset will map the room or area where you are, and you can start to play, build, work, and interact.

Regarding embedded extended reality, right now, we talk about extended reality as separate from other types of technology. It's new, it's cool, we're trying to figure it out, and we're innovating all the time. Call that Phase #1.

Phase #2 will be what I just described… extended reality as a set of common apps on your technology. You won't need to build the app itself; you'll just build inside the app. Let's call that Phase #2. It will be here shortly.

Beyond is Phase #3. Extended reality will be embedded into every aspect of technology (and our lives). As you watch TV, you'll be able to simply ask, "What's that building?" Your TV will pause the show and talk to you about the name of the building, when it was built, the architect, the height of the building, its location, and so on. You'll even be able to take a virtual tour of the building, before returning to the TV show. You'll be able to do the same when you're taking photos with your phone.

You'll stand in front of a smart mirror trying on clothes and you'll be able to see what you look like in clothes of a different color or with different shoes on. Yes, yes… mirrors are becoming a part of our technology tool set. A **smart mirror** is a mirror (duh) that displays your image and also extended reality content like augmented information, avatars, changes to your hair style or clothing, and so on. Check our smart mirrors at https://www.chattersource.com/smart-mirror/.

The Blurring of Augmented, Mixed, and Virtual Reality

There is already a lot of confusion and cross over among augmented, mixed, and virtual reality, especially in the popular press. For example, if you search on augmented reality applications, you'll actually find lots of mixed reality examples. Many extended reality headsets support applications for augmented, mixed, and virtual reality. The same is true for many extended reality development tools.

It's not really a big deal; the lines between augmented, mixed, and virtual reality are becoming quickly blurred. I think someday we'll simply use the single term extended reality.

The Death of Traditional Eyewear

The miniaturization of technology is an important driver, in the past, the present, and will continue into the future.

Someday, we won't need today's headsets for enjoying mixed and virtual reality. Capabilities to do so will be built into our eyewear, which will most definitely include sun glasses. We already have smart or AR-enabled ski goggles like RideOn and smart swim goggles like FINIS smart Goggle.

While we may never really see the death of traditional non-technology eyewear completely, it will certainly take a second place to smart eyewear.

> (Google tried this several years ago with Google Glass. It abandoned the personal version eventually but did release an Enterprise edition. Echo Frames... these are audio glasses that give you hands-free access to Alexa. No display or camera, so not XR-enabling.)

The Birth of Real Holograms

Without getting into any level of technical detail, holograms as we know them today are not real holograms. Real holograms are created from a recording of a light field, not from a traditional camera lens. Thus, real **holograms** are 3D images with depth that have been created using light beams, millions of them. Using panels of lights sitting opposite of each other, two light beams from each panel intersect in space and create a pixel, a tiny dot that you can see. You can change the color of the light beams emanating from each panel to create a different color at the intersection. Do this with enough pairs of millions of light beams and you get a real 3D hologram with depth.

Google's hologram initiative is called **Project Starline**. It is a booth-based holographic videoconferencing tool. Very cool… check it out at https://www.youtube.com/watch?v=Q13CishCKXY.

Real holograms are coming. Hmmm… *real hologram*, another oxymoron?

The Synergies of 4th Industrial Revolution Technologies

The biggest opportunities of the 4th industrial revolution lie at the intersection of the technologies. Metaverse, for example, uses extended reality such as avatars and holograms, cryptocurrency, and NFTs to build a more realistic and immersive experience.

As you innovate in the 4th industrial revolution, try to combine multiple emerging 4th industrial revolution technologies. The result is not additive but rather multiplicative. Big, big change, much better world.

EXTENDED REALITY BY THE NUMBERS

In 2019 there were 440 million users worldwide of augmented reality. That number is expected to grow to 1.7 billion users in 2024. [13]

CHAPTER 5

3D Printing

One Pair of Shoes for Life

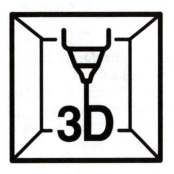

Question: Which of these has <u>not</u> been 3D printed?
  A. Toothbrush
  B. Doorstop
  C. Motorcycle
  D. Clothing
  E. Boat
  F. House
  G. Food
  H. Lamborghini

I'm sure you're getting good at these by now. You're right, all of the above have been 3D printed.

Basically, if it's in physical form, you can most likely 3D print it. And you may be thinking, "Sure, everything except living organisms." Think again... more on bioprinting in a moment.

3D printing is a subset of additive manufacturing. And it makes sense that additive manufacturing is the opposite of subtractive manufacturing. **Subtractive manufacturing** is simply taking something big and making it smaller. We've been doing some form of this since we invented tools. Early civilizations took a big rock and chiseled it down into a statue, bust, figurine, or part of a pillar. Today, we take trees and whittle them down to 2x4s, tooth picks, and matches.

Using the subtractive manufacturing approach, we slice, press, squeeze, cut away, whittle, etc. until we get the finished product we want.

Additive manufacturing is the opposite of subtractive manufacturing and includes things like cast molding, injection molding, and – of course – 3D printing. 3D printing has become so synonymous with additive manufacturing that we now use the terms interchangeably. In fact, if you go to Wikipedia and search additive manufacturing, you'll be redirected to the page on 3D printing.

3D printing is the construction of a 3-dimensional object by typically adding (i.e., printing) layer upon layer of liquid material until the object has been completely printed or constructed. The material is initially heated to a necessary liquid state and then printed. Once a layer cools, more material is heated and printed on top of the previous layer. Rinse and repeat until the object is completely printed.

▲ ▲ ▲

Interesting and Cool Innovations in 3D Printing

In the opening question for this chapter, I put up a bunch of stuff that has already been 3D printed. The mundane and boring include things like toothbrushes and doorstops. Boats are kind of interesting, check out https://www.youtube.com/watch?v=9U1UbMmNywA. But the others truly illustrate the power and disruptive impact of 3D printing. Let's explore.

I'll point you toward some videos and resources on each of these. Review them at your leisure.

Motorcycles

German company APWorks 3D printed an electric motorcycle. It weighs just 35 kilograms, about 1/10 the weight of a Harley Davidson. It can travel up to 80 kilometers per hour, almost 50 miles per hour. It was 3D printed using an aluminum alloy as the material. So, it's almost as light as aluminum but has the strength of titanium. Watch the video at https://www.youtube.com/watch?v=prErZtXFrSw&t=15s.

Imagine being able to design and print a motorcycle or bicycle to fit your exact body style. A seat that fits your posterior perfectly. The handle bars the perfect distance for your arm length. The pedals perfectly placed for your feet and the length of your legs. And you'll get to personalize your motorcycle or bike with your choice of colors, frame style, and accessories like a holder for your IoT-enabled water bottle.

Clothing

I think 3D printing is going to be big business in clothing. While we can't yet print purely in fabrics like cotton, we're getting close. Imagine designing and printing your own clothes at home. It will become a reality. You'll be able to add your own design features to create personalized clothing. You'll be able to tailor your clothing to your unique body style and dimensions. You'll be able to print in whatever colors you want. Take a look at these initiatives.

- Danit Peleg – this is my personal favorite, as I've been following Danit for several years now. She truly was a pioneer in the space of 3D-printed clothing. In 2018, she was named by Forbes as one of *Europe's 50 most influential women in tech*. Check her out at https://danitpeleg.com/ and also https://www.youtube.com/watch?v=3s94mIhCyt4.
- Ministry of Supply – has developed a 3D printing process for knits, specifically blazers for both men and women. Using its technique, Ministry of Supply has noted a 35% reduction in material usage. [1].
- Julia Daviy – and I quote her, "Filament is not ready to replace fabric completely just yet, but it's only a matter of time. As it stands today, the technology is already good enough to create better clothes than certain materials, like leather. For example, I created a top and a skirt that look as if I used a laser to intricately cut them from a piece of leather, but it's entirely flexible and biodegradable, vegetable based plastic. It was faster, cheaper, and more sustainable than using leather." [2] See what Julia is doing at https://juliadaviy.com/.
- Viptie 3D – 3D-printed neckties and bow ties for men. [3]

- Futurecraft 4D – an initiative by Adidas and Carbon to 3D print shoes. [4]
- 3D-printed jewelry – lots of great stuff going on in this space. See, for example, Sculpteo. [5]

I'm stopping here with the list but you should explore more own your own. It really doesn't matter the gender, the type of clothing, shoes, accessories, or even eyewear. [6] (Sorry, I had to throw in one more to look at. Now, go explore the rest on your own.) 3D printing has a place in all aspects of clothing and the fashion industry.

Think about reducing waste, creating clothing that is biodegradable, and addressing the big hairy audacious problem of fast fashion. I hope you're up to the challenge.

Again, lots of great opportunities in this space.

Houses

Homes are expensive, time-consuming to build, and require significant energy and resources.

But a 3D-printed home can be quickly built at a fraction of the cost of a traditionally-built home. And many 3D-printed homes are using waste and recycled material, making them eco-friendly.

- Chinese company WinSun has made a number of achievements in this area. It 3D printed a house in just one day. It also 3D printed a 5-story apartment complex and an 11,840 square-foot villa. All these printing efforts use industrial waste, recycle construction material, and cement. [7]
- The United State Marine Corps 3D printed a 500 square foot barracks made entirely of cement in just 40 hours. Normally, it would take 5 marines a full week to build the equivalent barracks using wood. [8]
- Construction company ICON has 3D printed homes all over world. One of its efforts was to 3D print homes in the Austin, TX area, which are now for sale. The homes range in size from two-bedroom to four-bedroom models. Each home took approximately 5 to 7 days to print. [9]

- Italian-based Mario Cucinella Architects, in conjunction with Italian additive manufacturing company WASP, is 3D printing homes with the primary source of material being the surrounding soil, coupled with water and rice husks. These 645 square-foot homes take about 200 hours to print and are dome shaped. [10]

That's just a few of the many initiatives for 3D printing homes. They're affordable, take minimal time to construct, and are very environmentally friendly when using recycled material, waste, and soil.

And, indeed, there are a number of other great initiatives in the housing space. Check out Boxabl (https://www.boxabl.com/) and any of the many different types of tiny homes (a good review is at https://www.countryliving.com/home-design/g1887/tiny-house/). As with 3D printing clothing, the 3D printing of homes can help address important problems such as safe and affordable housing.

(BTW, Elon Musk supposedly lives in a Boxabl.)

Food

The 3D printing of food will be fascinating to watch. Most people have no problem with 3D-printed clothing, homes, motorcycles, and the like. But when it comes to eating something that has been 3D printed, people start to hesitate. (Which is really odd considering that popular treats like Gummy Bears come from a form of injection molding, an additive manufacturing technology.)

But, I do believe that 3D-printed foods, especially those that we 3D print in our homes, will become a part of our everyday lives. You'll be able to add the vitamins, nutrients, and supplements you want. You'll be able to custom print mild, medium, or hot buffalo wings (from our Pudding on Demand discussion).

Consider these examples of 3D printing food. [11]

- Food Ink, a pop-up restaurant in London, 3D prints its entire food and dessert menu. Taking it one step further, Food Ink also 3D prints its chairs, lamps, and decorations.
- 3D-printing company Beehex provides the technologies for 3D printing pizzas and cookies. (NASA considered using this technology to make pizzas for astronauts.)
- Open-Meals is working to 3D print meals specifically designed for the nutritional needs of each patron. One of Open-Meal's initiatives is Sushi Singularity. When making a reservation, you provide biometric samples of yourself including your DNA. Singular Sushi will use that information to 3D print a meal tailored to your nutritional requirements.
- Novameat and Redefine Meat are 3D printing plant-based versions of meat products. They even use plant compounds that taste like the blood, fat, and muscle in traditional meat flavors. Once you buy them, just throw them on the grill and cook to your liking.
- Neatherlands-based Upprinting is collecting food typically wasted because of over-ripeness or "ugliness." It purees those foods to 3D print biscuits.
- LA-based Sugar Labs describes itself as, "rogue chefs, architects-turned-designers, and tech geeks" who 3D print sugar-based treats and desserts. (I wonder if they do pudding.)

Again, very short list of probably hundreds of initiatives in the 3D-printed food space.

Want to try it at home? Consider these food 3D printers: [12]
- Mycusini by German-based Print2Taste (www.mycusini.com)
- PancakeBot, a successful kickstarter campaign (www.pancakebot.com)
- Procusini also by Print2Taste (www.procusini.com)

Lamborghini

Okay, I had to include this one because it involves a 12-year-old young man and his father in my neck of the woods, Colorado. Xander Backus and his dad, Sterling, have been working for years to 3D print an exact working, road-ready replica of a Lamborghini. While you may think that Lamborghini doesn't want anyone "printing" its cars, Lamborghini has actually supported the father-son team and provided a 2-week loaner of a Lamborghini Aventador S. Check out one of the many videos about this initiative at https://www.youtube.com/watch?v=isru_Az3GGA&t=83s.

Bioprinting

Bioprinting (or **3D bioprinting**) is the use of 3D printing technologies, combined with special filaments such as bioink and other biomaterials, to replicate parts that imitate bones, natural tissues, tendons and ligaments, skin, blood vessels, and even organs.

Let's talk about 3D-printed synthetic living skin in burn treatments as an example.

More than 11 million people annually require burn-related medical procedures to improve the functional and aesthetic outcomes of burn wounds, Traditionally, burn treatments have included primary closure, burn wound excision with subsequent skin grafts, and skin substitutes. [13]

Now, medical professionals can 3D print synthetic living skin tissues. The print process combines the creation of a **scaffold** (polymeric biomaterials that provide the structural support for cell attachment and subsequent tissue development) with the depositing (3D printing) of living cells within the scaffold. The bioprinted tissue is allowed to mature and then used as a skin graft of sorts on the burn victim.

Other 3D bioprinting research and applications include:
- Breast implants [14]
- Bone tissue [15]
- Ears and noses [16]
- Blood vessels, nerves, and muscle tissue [17]

- Complex bone tissue [18]
- Related to 3D bioprinting are initiatives like e-NABLE, a global community of "Digital Humanitarian" volunteers who are working to 3D print upper limb prosthetic devices such as hands. The community is open-source, so anyone can use, modify, and reprint a design created by someone else. [19]

3D bioprinting is still very much in its infancy stages of research. But considerable progress is being made daily.

You should begin to think about the synergies of combining multiple 4th industrial revolution technologies. Consider 3D printing and IoT. In 2018, researchers at the University of Minnesota successfully proved they could 3D print biological sensors directly onto the skin of a human being. Fascinating application at the intersection of two 4th industrial revolution technologies. [20]

▲ ▲ ▲

3D PRINTING BY THE NUMBERS

The global 3D printing market is expected to grow from $12.6 billion in 2021 to $34.8 billion by 2026. [24]

Materials Used in 3D Printing – Filament and Resin

You can 3D print with just about any material or simulated material you want. Depending on your choice of different type of 3D printer, you'll either be printing with **filament** (a solid spool of material) or **resin** (a tank of liquid material). Filament and resin are essentially the same; they differ in their starting form (filament is solid while resin is already liquid) and the process of heating and printing them.

Either way, the common materials include:
- PLA (polylactic acid) - the most common in home printing, the printing of prototypes, etc. Very inexpensive but also not of high quality or strength.
- ABS (acrylonitrile butadiene styrene) - the second most commonly used, and not much different from PLA.
- PET (polyethylene terephthalate) - common clear plastic like in water bottles.
- Nylon - synthetic polymer; very strong and durable.
- TPE (thermoplastic elastomer) - think flexibility like in the soles of running shoes.
- PC (polycarbonate) - the strongest; think bicycle helmets.

You can also use other filaments/resins, what we call "exotics." Some of those include:
- Hemp
- Wood
- Biodegradable
- Conductive
- Glow in the dark
- Magnetic
- Ceramic
- PVA (polyvinyl alcohol) (dissolves in water, the clear plastic material around a laundry detergent or dish washer pod)
- Wax

If you go the filament route, materials come on a spool, about 6 inches in diameter and weighing a couple of pounds. Prices range from $20 per spool for the cheaper filaments (PLA, ABS, etc.) up to $50+ for the exotics. If you go the resin route, you buy resin by the liter, with cheaper resins costing about $40. When you get a liter of resin, you pour it into a tank that your 3D printer uses. The tank itself costs about $80 and you'll have to replace the tank periodically.

▲ ▲ ▲

The 3D Print Process

The 3D printing process includes 4 steps:
1. Ideate and design on paper
2. Use 3D modeling software to create the digital model
3. Use slicing software to create the instructions your 3D printer needs to print the object
4. Print the object using your 3D printer

Of course, this is a very iterative process. So, when you're first starting on a new object, I recommend that you use PLA or ABS material, the cheapest of the materials. You'll undoubtedly print your first try at the object and then decide to make some changes. Using PLA or ABS to get to a final design is the cheapest and often fastest way to go (the more exotic the material, the longer it usually takes to print).

In the first step, take some time to design your object on paper, thinking about the various dimensions, curvatures, etc. If you want to 3D print a copy of an object you already have (either exact replica or modified version of it), you can take a photo of it and load it into your modeling software.

In the second step, you'll use **3D modeling software** to create a 3D CAD (computer-aided design) digital model of your object. The end result of this step is called an **STL file** (**Standard Tessellation Language** or **StereoLithography file**), a digital file that describes the surface geometry of your object. It's rather like using Word to create a .doc or .docx file.

This is the step in which you'll spend most of your time. You can start with a blank slate and use your 3D modeling software to create your model, you can start with a photo as I just mentioned, or you can visit any number of sites and download an STL file of the object you want to print. If you go the latter route, you can modify the design to fit your needs.

There are some really great sites that offer free 3D CAD designs, including [21]:
- Cult (www.cults3d.com)
- Free3D (www.free3d.com)
- GrabCAD (www.grabcad.com)
- MyMiniFactory (www.myminifactory.com)
- Pinshape (www.pinshape.com)
- STLFinder (www.stlfinder.com)
- Sketchfab (www.sketchfab.com)
- Thingiverse (www.thingiverse.com)
- Yeggi (www.yeggi.com)
- YouImagine (www.youimagine.com)
- 3DShook (www.3dshook.com)
- 3DWarehouse (www.3dwarehouse.sketchup.com)

Many of these support a very active community of 3D designers with whom you can collaborate on design.

As for the 3D modeling software itself, there are many options, and some have free personal licenses and/or free licenses for students. There is an important consideration in choosing your 3D modeling software. 3D modeling software is used for a variety of purposes, not solely for 3D printing. For example, you can use 3D modeling software for creating computer graphics for applications like games, virtual reality, and mixed reality. 3D modeling software for 3D printing is often referred to as *solid modeling software*.

You need solid modeling software that creates call *manifold* or *water tight*. With a manifold model, the walls have some thickness, and that is necessary for 3D printing. Some 3D modeling software creates walls that have zero thickness, the types of walls that are used in computer graphics and games. This type of 3D modeling software is called *polygon modeling software*. If you use polygon modeling software, you can still create 3D-printable STL files, but it takes a few more steps.

Below are some modeling software options for 3D printing. [22]

- Tinkercad (www.tinkercad.com)
- Blender (www.blender.org)
- BRL-CAD (www.brlcad.org)
- DesignSpark Mechanical (www.rs-online.com)
- FreeCAD (www.freecadweb.org)
- OpenSCAD (www.openscad.org)
- Wings3D (www.wings3d.com)
- 3D Slash ((www.3dslash.net)
- SketchUp (www.sketchup.com)
- Fusion 360 (www.autodesk.com)
- MoI 3D (www.moi3d.com)
- Rhino3D (www.rhino3d.com)
- Cinema 4D (www.maxon.net)
- SolidWorks (www.solidworks.com)
- Maya (www.autodesk.com)
- 3DS Max (www.autodesk.com)
- Inventor (www.autodesk.com)

If you want my recommendation, I'd start with Tinkercad, Blender, 3D Slash, or SketchUp. These are very easy tools to learn and offer a free version.

If you love this space and want to pursue a career in the 3D modeling (and printing), you should really consider the suite of 3D modeling tools offered by Autodesk, including Fusion 360, Maya, 3DS Max, and Inventor. Fusion 360 is a powerful, general-purpose 3D modeling tool that is very popular in the industry. Maya is great for developing characters. 3DS Max is great for games and the entertainment space. Inventor is great for mechanical design. As these all come from Autodesk, so the interface is similar.

Once you've built your model and resulting STL file, you've got one more step to go before 3D printing. In this 3rd step, you need to use slicing software. **Slicing software** takes an STL file (the surface geometry of your object) and slices it to create the layers needed for 3D printing. The result is **g-code**, a CNC (computer numerical control) set of instructions that basically tells your 3D printer how to print your object layer by layer.

Again, lots of options here, including: [23]
- Cura ((www.ultimaker.com)
- Simplicity3D (www.simplify3d.com)
- Slic3r (www.slic3r.org)
- Repetier (www.repetier.com)
- KISSlicer (www.kisslicer.com)
- IdeaMaker (www.raise3d.com)
- OctoPrint (www.octoprint.com)

Most of these work with most of the popular 3D printers that are out there. When choosing slicing software, do make sure that your 3D printer can work with the slicer you're considering.

As for my recommendation, start with Cura, Simplicity3D, Slic3r, or KISSlicer. Free versions and easy to learn and use.

You'll also need to consider if you want a 3D printer that can print multiple materials simultaneously. Basic versions of some slicing software will work with only one material. You'll have to upgrade (and probably pay for it) to work with multiple materials.

▲ ▲ ▲

Selecting a 3D Printer

There are 2 dominant types of 3D printers.

1. ***Fused deposition modeling (FDM)***, in which a spool of solid filament is heated and printed by a heated printer head called an extruder. FDM are the most common type of 3D printer. Because FDM 3D printers work with material called filament, they are often referred to as filament 3D printers in addition to FDM 3D printers.

2. ***Digital Light Processing (DLP)***, in which liquid material called resin is heated and then printed in similar fashion to FDM 3D printing. Because DLP 3D printers work with material called resin, they are often referred to as resin 3D printers in addition to DLP 3D printers.

As you can see in the table below, there are definitely differences between filament and resin 3D printers. Resin 3D printers, overall, are on the higher price side, including cost of the printer, cost of the material, and cost of maintenance and upkeep. However, resin 3D printers yield a much higher quality finished product. Because the resin is already in liquid form, resin 3D printers also tend to be faster than filament 3D printers.

	FDM or Filament 3D Printer	DLP or Resin 3D Printer
Printer Cost	Lower	Higher
Material Cost	Lower	Higher
Maintenance Cost	Lower	Higher
Print Speed	Slower	Faster
Quality of Finished Product	Lower	Higher

Most professionals will use both. They will use a filament 3D printer to do prototyping and then a resin 3D printer for the final quality print. If you're just starting out in this space, I recommend you buy a filament 3D printer first, a nice inexpensive option to get started.

Choosing a 3D printer is similar to choosing a regular paper printer… what capabilities do you need, how fast do you want to print, in what sizes do you want to print, do you need single or multiple colors, what's your budget? These are all important questions. In the 3D printing realm, they're a bit more complicated than a regular paper printer.

So, let's try to break this down as simply as possible, focusing on personal 3D printing needs.

- Price: They range in price from a few hundred to a few thousand dollars. As I noted before, resin 3D printers are definitely more expensive than filament printers.
- Common/Exotic material question: Basic 3D printers work with only basic materials like the commons listed previously. If you want to work with exotic materials, you'll need a more "advanced" 3D printer and one that costs more.
- Multiple material question: Basic 3D printers will only one print from one material at a time. If you want to embed (i.e., print) magnets into your printed products because you're making wild and crazy refrigerator magnets, you'll need a more expensive printer that can simultaneously print from multiple filament spools or resin tanks. Some 3D printers can print from more than 2 filament spools or resin tanks; these are obviously more expensive. This also includes multiple colors of the same material.
- Build area: **Build area** addresses the maximum size of the object you can 3D print. Smaller, more inexpensive 3D printers have smaller build areas than larger and more expensive 3D printers that have larger build areas. At the time I wrote this, typical sizes for build areas included: width of 3 to 12 inches, depth of 3 to 8 inches, and height of 5 to 10 inches. The more width, depth, and height you want, the more expensive the printer.

▲ ▲ ▲

3D Printing Impact on Sustainability: From Big R to Small r

Reduce-Reuse-Recycle… I love it. Many of our big hairy audacious problems such as fast fashion, food waste, and the use of fossil fuels are antithetical to the notions of Reduce-Reuse-Recycle. And it takes BHAT (big hairy audacious thinking) to address these BHAPs (big hairy audacious problems).

I refer to Reduce-Reuse-Recycle as the big Rs. They're big because of their importance and because we typically address them at the macro level. Consider recyclable plastic soda or water bottles. They appear in mass quantities on grocery store shelves. We consume their contents in mass quantities. Then, hopefully, we throw them into blue recycle bins, and the recycle bins are hauled off by trucks to a recycle facility, where they are repurposed for use again. Rinse and repeat.

That's good and I have no problem with it. But what if we could reduce-reuse-recycle at the micro level, in our own homes? 3D reprinting offers us that opportunity.

3D reprinting uses existing items to recycle filament/resin and use that recycled filament/resin to 3D print new versions of the items. I'll illustrate with a simple example, toothbrushes.

I think someday we'll 3D print our own toothbrushes. Really. Visit a website such as Thingiverse (www.thingiverse.com), find the toothbrush you like, download the STL file of it, and print all the toothbrushes you want. The cost? Maybe you'll have to pay $.99 for the download. (I actually think they'll  be free. Thingiverse designs currently are free.) When you need a new toothbrush, the material will probably costs less than 10 cents. You can have a new toothbrush every month of the year for about a dollar.

But what to do with the old toothbrushes? The solution is 3D reprinting, and it's already possible.

As an example, the ReDeTech Protocycler+ allows you to recycle your old 3D-printed products and use the recycled filament of those to make new products. [25] So, when you're done with your current toothbrush, use the filament in that toothbrush to 3D reprint a new toothbrush. You don't have to recycle your old toothbrush at the macro big "R" level. You simply do it in your home, small "r" level.

In the fall of 2018, when we rolled out the 4th Industrial Revolution course, the required intro course for all business majors and minors at the University of Denver, a wonderful group of students envisioned a future possibility of this.

Their notion was **one pair of shoes for life**. When a child is born, parents can have a 3D-printed pair of shoes custom made for the child. Of course, the shoes won't last long, because the child grows. When the shoes no longer fit, return the shoes, pick out a new style and size, and the company will use the returned shoes to 3D reprint a new pair of shoes. At the age of 18, when the parents order a new pair of shoes (and return the old ones), the child has a pair of shoes that are made from the exact same material of the shoes he/she/they had at birth. What a beautiful story, and what a powerful example of small "r" reduce-reuse-recycle.

I think this happens with just about anything you care to name. Kitchen utensils – like spatulas, whisks, and stirring spoons – get old, worn out, or break. Throw them in your recycler and make new utensils from them. Break a shovel, make a shovel. Your favorite shirt gets old and worn out, make a new shirt from it.

Think about candles. (I'm a candle lover.) When a candle burns out, you throw away the remaining wax. Why not use the wax of the old candle to make a new candle? All you have to do is use the wax of the old candle to 3D reprint a new candle. Add the wick, which you can easily do with a 3D printer that works with multiple filaments/resins, and you have a new candle. You'll also be able to easily change the scent and color of the new candle.

BTW, wax already exists as an exotic material for 3D printing.

To truly understand the significance of 3D reprinting, think about interstellar space travel. (It's closer than you think.) When something breaks or wears out on a space ship, you're not going to go out to the shed or supply room and find a replacement. You can't carry a backup of everything on a space ship. Instead, you'll use 3D reprinting to use the filament of the old/broken item to make a new item.

▲ ▲ ▲

3D PRINTING BY THE NUMBERS

The number of manufacturers using additive manufacturing for full-scale production doubled from 2018 to 2019. [26]

IMAGINE FORWARD

3D Printing Everything

Everything seems to be a common thread in our predictions of the future. And it's certainly applicable in 3D printing. It's hard to imagine what we won't 3D print in the future. Perhaps big, huge pieces of furniture but, then again, we're already 3D printing entire homes. Why not furniture?

DIY becomes PIY

Do-it-yourself meets 3D printing and becomes print-it-yourself. 3D printers will become so easy to use that you won't need to hire someone to design and print your fill-in-the-blank; you'll just do print-it-yourself.

Recycle-Reprint-Print

According to the EPA, the U.S. recycled 46 million tons of paper in 2018. [27] Let's recycle it at home and not just produce more paper, but produce paper that is automatically fed into a regular printer when we need to print on paper. Print-on-demand, paper-on-demand.

3D Printing Fake Fingernails or Directly on Fingernails

It's going to happen. The 3D printing of fake fingernails already has some limited success. 3D printing on the human body has as well. Just a matter of time before a 3D printer applies your fingernail polish.

3d Printing Makeup on Your Face

See above and substitute makeup for fingernail polish.

Edible Packaging

I love this one. Think of the amount of trash and recycle we would eliminate if food came in 3D-printed edible packaging. Package of spaghetti noodles… just drop the whole thing – the noodles and the packaging – in hot water. Drink a bottle of soda, eat the bottle the soda came in. Eat a candy bar without ever taking off the wrapper. (Don't believe me… send me a notable quotable.)

3D PRINTING BY THE NUMBERS

In 2018, 1.42 million 3D printers were sold worldwide. That number is expected to increase to 8.04 million by 2027. [28]

CHAPTER 6

Autonomous Vehicles

The Death of the Driver's License

Time for you to answer another question. According to the National Highway Traffic Safety Administration (NHTSA), what percent of automobile accidents in the U.S. are caused by human error? [1]
  A. 54%
  B. 64%
  C. 74%
  D. 84%
  E. 94%
  F. 99%

May have fooled you here. The answer is E, 94% of all automobile accidents in the U.S. are caused by human error. The "errors" include things like excessive speeding, distraction, incorrect assumptions about what other drivers are going to do, tailgating, and not checking traffic lights before pulling out or changing lanes. [2]
  - Over 6 million automobile accidents happen each year in the U.S., one accident every 5 seconds. [3]
  - In 2020, 38,680 people in the U.S. died in automobile accidents. [4]
  - According to the NHTSA, 10,142 people died in drunk-driving crashes in the U.S. in 2019. That was the lowest number of drunk-driving related deaths since 1982, but still meant that one person died every 52 minutes in a drunk-driving accident. [5]

Grim statistics.

As one of the 7 core 4th industrial revolution technologies, we hope and believe that autonomous vehicles can reduce automobile accidents and deaths. Additionally, autonomous vehicles will enable significant innovation and industry disruption.

An **autonomous vehicle** is a vehicle that can guide itself without human conduction, using sensors, software, and a variety of driver-assistance technologies to navigate the vehicle. Driver-assistance technologies include things like adaptive cruise control, remote parking, and lane departure warnings.

When we think about autonomous vehicles, we most commonly do so in terms of personal automobiles or cars. But *vehicle* can include long-haul trucks, heavy equipment vehicles, airplanes, motorcycles, and any other type of vehicle that moves people, cargo, and the like.

Here, I'll be talking with you about autonomous vehicles mainly within the context of personal automobiles and cars. (You can easily extrapolate to the other types of vehicles.)

▲ ▲ ▲

The Range of Autonomy from Nothing to Everything

The range of autonomy goes from none to completely autonomous, although we are not yet at the fully autonomous level.

According the Society of Automotive Engineers there are 6 levels of driver-assistance technologies (i.e., autonomous vehicles). (See Figure 6.1) [6]

- Level 0 – no automation at all; the driver does all the driving. At this level, you're getting no support, or perhaps very limited support only in the form of warnings like lane departure and blind spots.
- Level 1 – some support for steering (e.g., lane centering) or braking and accelerating (e.g., adaptive cruise control).
- Level 2 – some support like lane centering (steering) and braking and accelerating (adaptive cruise control) at the same time.
- Level 3 – the car can drive itself under certain limited conditions. The car may require that you take control of driving.
- Level 4 – same as #3 (driving under limited conditions), except that the car will not require that you take control of driving.
- Level 5 – the car can drive itself everywhere under all conditions.

Figure 6.1 **Levels of Autonomy in Autonomous Vehicles**

LEVEL 0	LEVELS 1 & 2		LEVEL 3	LEVELS 4 & 5	
You control the car completely.	You are mostly in control of the car and the car requires your constant supervision.		The car can completely control itself under limited conditions.	The car can completely control itself. You will not be required to take back control of the car.	
No driver-assistance technology.	Some driver-assistance technologies.				
	Steering **OR** braking assistance, but not both.	Steering **AND** braking assistance.	You may be required to take back control.	Under limited conditions.	Under **ALL** conditions.
You may get warnings for things like lane departure.					

It's important to understand that <u>you are the driver completely</u> in the first 3 levels (0 through 2). At no time will your car take complete control of all driving functions. You must be in the driver's seat at all times, ready to take total control of your car at any time. In level 5, alternatively, there may not even be things like a steering wheel and pedals present in the car. Won't that be wild to see? Levels 3 and 4 are obviously somewhere in the middle.

▲ ▲ ▲

AUTONOMOUS VEHICLES BY THE NUMBERS

75% of all cars worldwide will be fully autonomous by 2040. [7]

Most Common
Driver-Assistance Technologies

Driver-assistance technologies control a specific aspect of your car's operation, mostly in steering, braking, etc. Technically, warnings for things like lane departure and driver fatigue are not driver-assistance technologies. They warn you but they do not take corrective actions themselves. However, you will see these types of warnings listed under driver-assistance technologies. (The list below includes them.)

Driver-assistance technologies include:
- Lane departure warning - usually a beeping sound when your car notices you are "drifting." Some vehicles also provide haptic feedback by making the steering wheel vibrate.
- Driver fatigue warning – usually a beeping sound when your car notices that you seem to be steering in a less-than-optimal fashion (e.g., weaving within your lane).
- Blind spot warning – usually a beeping sound when you're moving in the direction of an object or an object is moving toward you.
- Automatic braking (collision avoidance) – when objects suddenly appear in front of the vehicle.
- Adaptive cruise control – set your speed and the distance you want to stay behind cars in front of you. Your car will slow down when it gets within a certain distance of the car in front of it.
- Parallel parking – this certainly seems to be a lost art. No explanation necessary.
- Weather-related assistance – controlling wind shield wipers based on moisture, moving between dim and bright lights based on fog/rain/snow, shifting into or out of all-wheel or 4-wheel drive based on snow and ice.
- Remote parking – while not even being in your car, your car navigates getting into and out of tight parking spots.
- Hands-free navigation/steering – no explanation necessary.

"Limited Conditions"

The term *limited conditions* applies to levels #3 and #4. In those levels your car can drive itself – including steering, accelerating, and braking.

But your car will only be able to engage all those driver-assistance features if certain conditions are met.

Limited conditions already do and may in the future include:
- Highway driving only (not on busy city streets)
- (Good) weather
- (Minimal or no) traffic congestion
- (No) construction areas in which speed is limited

Those are the major ones, and they make sense. A long stretch of highway in good weather with little other traffic and no construction is optimal (right now) for an autonomous vehicle to fully take control of steering, accelerating, and braking.

▲ ▲ ▲

AUTONOMOUS VEHICLES BY THE NUMBERS

2033 will mark the end of the manufacturing of consumer vehicles with no driver-assistance technologies. [8]

Autonomous Vehicles as a Combination of Other 4th Industrial Revolution Technologies

As a function of the other 4th industrial revolution technologies, autonomous vehicles look like this:

$$AV = AI + ST + CT + IoT (+Ba)$$

Autonomous vehicles, as a core technology in the 4th industrial revolution, are a combination of artificial intelligence, sensing technologies, communications technologies, and IoT. I also included battery, as I believe all vehicles in the future will be fully electric.

Artificial Intelligence in Autonomous Vehicles

Artificial intelligence (AI) is key for autonomous vehicles. Autonomous vehicles must be intelligent and continue to learn even more as you drive (or the car drives itself).

- Preloaded Intelligence – In the section on AI, we talked about how AI learns through training. When you buy an autonomous vehicle, it will come to you having already driven perhaps millions of miles in a simulator. So, it will already know how to recognize and change lines, how to accelerate and brake based on the traffic around it, recognize and interpret traffic signs, what to do at an intersection based on the lights and surrounding traffic, how to respond to the presence of pedestrians, cyclists, and so on.
- Continual Learning – As you use your autonomous vehicle (or allow it to drive itself), it will continue to learn. For example, it will learn more about your neighborhood, your garage for parking, and the peak and lull times for traffic getting in and out of your neighborhood.
- Collaborative Decision Making – Autonomous vehicles will communicate with each and decide what's best for the collective of vehicles on the road and in proximity.

Sensing Technologies in Autonomous Vehicles

Just as you respond to hearing, seeing, and feeling (omitting smelling for obvious reasons) while driving, your autonomous vehicle will do the same using a variety of hearing, seeing, and feeling technologies. (I would encourage you to read Chapter 9 on Sensing Technologies.)

"Hearing" can include sirens of emergency vehicles and the backing-up beeping of maintenance vehicles. "Feeling" can include wind and road conditions (pot holes, ice, rain, etc.).

"Seeing" is critically important to an autonomous vehicle, just as it is to you while you drive. To see , autonomous vehicles use a combination of sonar, radar, and Lidar. Autonomous vehicles also use a variety of cameras – including 360 cameras – for seeing.

Sonar, Radar, and Lidar in Autonomous Vehicles

All three of these technologies are used in various forms by autonomous vehicles. Depending on the manufacturer, you will see a different level of focus for each technology. Elon Musk, for example, is a stronger advocate of sonar and radar as opposed to Lidar.

- Waymo (Google/Alphabet subsidiary) – uses Lidar as its main seeing technology. It also uses radar to help detect and interpret objects in rough weather conditions (fog, snow, etc.).
- Aurora Technologies (previously Uber's self-driving car initiative) – uses primarily both radar and Lidar.
- Tesla – uses primarily sonar and radar. Most Tesla vehicles have 12 ultrasonic sensor units, coupled with radar capabilities to detect objects up to 500 feet away.

Regardless, it takes multiple seeing technologies for autonomous vehicles to be effective in driving without human conduction. Radar, for example, can detect objects up to 60 miles away. That's helpful for changing routes based on anticipated traffic congestion and accidents. But, Lidar has a much higher degree of accuracy for detecting and identifying objects that are closer. That's helpful for determining if a stationary object is a sign post, a pedestrian, or a tree.

Communications Technologies in Autonomous Vehicles

When most people think about autonomous vehicles, they think about their own car driving itself around. That's a very narrow view. It may be true right now, but it won't be in the future.

As we get more autonomous vehicles on the road, they will communicate with each other to optimize driving. Examples:

- Once a traffic light turns green, all waiting vehicles will simultaneously move in tandem through the intersection. No delay because people aren't paying attention.
- When approaching a construction area with reduced lanes, all vehicles will communicate to determine the optimal way to combine lanes while maintaining speed. (Won't that be nice?)
- If your vehicle is ahead of mine, my vehicle will see what your vehicle sees, for example, a pothole.

Vehicles-to-vehicles communication is so important for autonomous vehicles. It will certainly provide for the optimization of driving in terms of time, speed, and other forms of efficiency. But, it will also save lives. Right now, when a driver suddenly slams on the brakes, there is the potential for being rear-ended because of the time it takes the driver  behind to notice and respond. In the future, an autonomous vehicle will constantly inform other vehicles of what it is doing.

IoT in Autonomous Vehicles

Just about everything we've discussed for sensing and communications technologies will be in the form of sensors and implemented via IoT. Camera, Lidar, Bluetooth, WiFi, and many other sensing and communications technologies will be *a network of Internet-connected objects that collect, process, and exchange data* (our definition of IoT from Chapter 1).

▲ ▲ ▲

Leading Companies in Autonomous Vehicles

As you can imagine, all the long-established car companies are working in the autonomous vehicle space… Toyota, Volkswagen, Daimler, Ford, Honda, BMW, GM, Mercedes-Benz… everyone. There are also many new players. Some are well-known like Tesla and Waymo, while a few names may be new to you.

The country of China is an interesting player, of sorts, in this market. While most other countries are researching and operating internationally in the autonomous vehicle space, China isn't. China-based companies include Pony.ai, AutoX, WeRide, Baidu, and DidiChuxing.

Consumer Autonomous Vehicles

The leading companies in the consumer autonomous vehicle space include:
- Tesla
- Apple
- Kia-Hyundai
- Ford
- Audi
- Huawei

You don't hear much about Apple, but it's working aggressively to deliver a completely autonomous vehicle (level #5) by 2025. It has no plans to produce autonomous vehicles at any other level, just fully autonomous.

Robotaxi Autonomous Vehicles

Robotaxis are exactly what the name implies, autonomous vehicle taxicabs. Waymo is the most well-known in this segment. Others include:

- Cruise (joint venture of Honda, GM, and a few others)
- Argo (Ford and Volkswagen)
- Motional (Hyundai)
- Zoox (Amazon)
- Aurora (previously Uber's self-driving initiative)

Long-Haul Truck Autonomous Vehicles

This space is interesting in that the transportation model being pursued by many companies is quite different from what most people think about. Those companies envision autonomous long-haul trucks only operating on the highways. Let's consider the Denver to Salt Lake City run.

There would be a transportation hub just outside of each of these cities. Human-driven trucks originating in Denver, for example, would drive a trailer to the transportation hub. The trailer would then be connected to an autonomous long-haul truck  which would drive it to the hub just outside of Salt Lake City. There, the trailer would be connected to a human-driven truck that would take it into the interior of Salt Lake City. In this way, the autonomous driving would occur only on highways.

Companies in this space include Kodiak, Embark, TuSimple, Waymo, and Aurora. The latter 2 are working in both this space and the robotaxi space.

▲ ▲ ▲

The Downsides to Autonomous Vehicles: It Can't All Be Roses

Every coin has 2 sides. Every rose has its thorns. Say it however you want, but there are always goods and bads to everything. Autonomous vehicles are the same.

- **Driver Job Loss** – most notably among professional drivers... taxi cab drivers, long-haul truck drivers, delivery drivers, and so on.
- **Unintended Consequence #1** – because autonomous vehicles will reduce accidents, they will have a (seemingly) adverse impact on the automobile insurance industry, autobody shops, the tow truck industry, and so on.
- **Unintended Consequence #2** – the majority of organ donations come from victims of automobile accidents. The wait time and list for organ donation are already long. They will get longer with fewer organ donations coming from victims of automobile accidents. (Perhaps bioprinting can help. See Chapter 5 on 3D Printing, specifically bioprinting.)

There are undoubtedly many other "downsides" to autonomous vehicles. We have to carefully weigh all the advantages and disadvantages and make decisions in the best interest of society as a whole.

▲ ▲ ▲

AUTONOMOUS VEHICLES BY THE NUMBERS

$37 billion – the projected size of the global autonomous vehicle market in 2023, up from $27 billion in 2021.
[9]

What You Can Do; What Your Car Can Do without You

You Can Do Anything

Once we get to fully autonomous vehicles (Level #5), you'll be able to do just about anything you want while your car drives you around (within the law, of course). Automobile manufacturers are already rethinking the interior design of a car. Could it have a bed so you can nap on the way to work? Will the notion of front seat and back seat go away, in favor of 4 chairs that face into the middle with a work table in the center?

Your Car Can Do Anything Without You

This will be interesting to see how it plays out. Certainly in your lifetime, your autonomous vehicle will drive you to work and then drive off to run errands, be an Uber driver, and go back to your house and take the kids to school. Your autonomous vehicle won't need you or anyone else as a passenger to go places and do things. There will obviously be much debate and legislation around this issue.

▲ ▲ ▲

AUTONOMOUS VEHICLES BY THE NUMBERS

Automotive Lidar shipments are expected to be 34 million units in 2032, up from 1.4 million units in 2022. [10]

IMAGINE FORWARD

The Death of the Driver's License

Someday, autonomous vehicles will not have a steering wheel, pedals, or any of the other gadgetry that is necessary for you to drive the vehicle.

When that happens, what's the point of having a driver's license?

The Death of Stop Signs, Traffic Lights, and Lanes

It's going to be so much fun to see what happens in this space. Think 20 years from now. I truly believe that all vehicles rolling off the manufacturing floor will be completely autonomous (Level #5). In fact, 20 years from now you may have to get a special permit to drive a non-autonomous vehicle on the road. In that future, the "certain conditions" clause will apply to non-autonomous vehicles, not the reverse as it is now.

At that point, things like stop signs, traffic lights, and lanes will be non-existent. Just think about this with me for a moment. Why do we have stop signs at an intersection? Simple, we need them so drivers can navigate without causing harm. In the future, all autonomous vehicles will approach an intersection and "negotiate" will all the other autonomous vehicles concerning the ordering of proceeding through the intersection. There won't be a physical sign telling your autonomous vehicle to stop. Using all the sensing and communications technologies, your autonomous vehicle will know that it can move on through the intersection because no other traffic is approaching the intersection.

In the same fashion, lanes won't be necessary. We've create lanes so people know how to position their cars. In the future, our autonomous vehicles will decide that. From east to west, for example, the collective of autonomous vehicles may decide to have 5 "lanes" of traffic going east in the morning and only 3 "lanes" of traffic going east in the afternoon. And that may change from day to day.

How many times and how much time have you wasted sitting at an intersection with a red light while no cars are traveling perpendicular to you? That won't happen in the future.

The Death of Review Mirrors

We need them so we can see behind our own car while changing lanes, parking, etc. Autonomous vehicles will have cameras.

The Death of Windshield Wipers

We need them so we can see out the front wind shield. Autonomous vehicles won't need to see through the front windshield.

The Death of Headlights and Turn Signals

I think you're beginning to get the idea. Today's vehicles have lots of stuff to help us drive. Autonomous vehicles won't need things like headlights and turn signals. Think about the typical car today. What won't be needed in the future?

Autonomous Vehicle Impact on Ride Share

So much disruption is going to occur in the ride share space over the next several years. The likes of the taxi cab industry, Uber, and Lyft will undergo dramatic changes as vehicles no longer require a driver.

Moreover, we'll begin to see "share" more than "own." Think about this possibility. You're considering a number of different apartment complexes to move into. You have to think about location, price, size, and amenities. Those "amenities" may very well include the size of the apartment complex's autonomous vehicle fleet, how many free minutes you get each month included in your rent payment, and how many times each month you can use an autonomous vehicle during peak times.

HOAs (home owners' associations) may offer this amenity for neighborhoods. Large retirement centers will undoubtedly do the same.

Mobile Shopping

Imagine a future with mobile shopping vehicles traversing through neighborhoods. During wintery months, the vehicles travel around with an inventory of snow shovels, gloves, ice scrapers, and salt. You'll use an app to "hail" such a vehicle to your location. Once it arrives, you go in, find what you want, and pay using facial recognition software.

This will become a profitable business model because the vehicle won't have a driver. It can run 24x7, needing only to "take a break" to replenish inventory and get a battery charge.

Your Autonomous Vehicle Will Pay for the Toll Road

Think combining autonomous vehicles with other 4th industrial revolution technologies, like cryptocurrency. You'll travel down a toll road and your autonomous vehicle will use cryptocurrency to pay the toll. Guess what, your autonomous vehicle will be connected to your bank account(s).

Your Autonomous Vehicle Will Pay for Parking

See above and substitute "parking" for toll road.

Your Autonomous Vehicle Will Pay for Car Washes

See above and substitute... well, you know what I mean.

Disruption to Supply Chain Management and Delivery Activities

Coupled with drones (our next chapter), autonomous vehicles will most certainly redefine supply chain management and delivery activities.

Delivery vehicles will perform their tasks 24x7, like the mobile shopping vehicles we just discussed. UPS, FedEx, and Amazon are already exploring delivery vehicles equipped with drones. The delivery vehicle itself may never stop, instead just drive through a neighborhood with drones launching from the top, dropping packages at homes, and then flying to the location of the delivery vehicle to get the next payload.

When this happens (and I believe it will), supply chain management goes from a 2-dimensional optimization problem to a 3-dimensional one. More on this in the next chapter on drones.

Out with the New, In with the Old

Someday, we won't call them autonomous vehicles. We'll simply call them *cars*.

AUTONOMOUS VEHICLES BY THE NUMBERS

The average American spends 17,600 minutes driving annually. That's just shy of 300 hours per year. [11]

CHAPTER 7

Drones

Male Bee, Bagpipe, or Flying Vehicle?

Time for another question. How would you define a drone?
  A. Flying vehicle
  B. Remoted controlled aircraft
  C. Unmanned aircraft
  D. Male bee from an unfertilized egg
  E. Musical note or cord
  F. Bagpipe
  G. Bladder fiddle
  H. Type of minimalistic musical style

Trick question of sorts (again), as all of the answers are correct. Drones, as we'll discuss them here, run the full gambit across many perspectives including aerial to underwater, commercial to personal, remote-controlled to autonomous, ad nauseum. That makes it difficult to provide any sort of succinct encompassing definition. Let's just say **drones** are flying vehicles without needing a pilot in the vehicle. (That still isn't completely correct as remote-controlled underwater vehicles are often called underwater drones.) They may be remote-controlled or completely autonomous. You may or may not need a "license" to fly one. Drones, what a term. (And what in the world is a bladder fiddle?)

▲ ▲ ▲

DRONES BY THE NUMBERS

The global drone market is expected to more than double from $18.8 billion in 2020 to $40.9 billion in 2027. [1]

Quick Lesson in Aviation

Before we launch (yes, pun intended) into our discussion of drones, let's have a quick aviation lesson to get a few terms in place.

- *Fixed wing aircraft* generate forward thrust. Commercial airline flights you take are in a fixed wing aircraft. The speed of the aircraft, the flow of air, and the shape and tilt of the wings generate lift, causing flight.
- *Rotary wing aircraft* generate vertical thrust. Just think about a typical helicopter. Enough said.
- *Single engine aircraft* have a single engine for providing thrust. For fixed wing aircraft, there is a single engine that propels the aircraft forward. For a rotary wing aircraft, there is a single engine that spins a single rotary blade.
- *Multi-engine aircraft* have multiple engines for providing thrust (duh). For fixed wing aircraft, there are multiple engines that propel the aircraft forward. For a rotary wing aircraft, there are multiple engines that each spin a different rotary blade. The latter type of aircraft is often called a *multicopter*.
- *Unmanned aerial vehicle (UAV)* – aircraft without any pilot, crew, or passengers aboard but is remotely piloted. Think personal drone. You may also see *remotely piloted aircraft (RPA)* used interchangeably with UAV.
- *Autonomous drone* – aircraft with the necessary software, sensors, and artificial intelligence to fly itself without human conduction.

▲▲▲

Consumer Versus Commercial Drones

There are significant differences between consumer (personal) and commercial drones.

Consumer Drones

Consumer drones are for your own use and enjoyment. You use a set of remote controls to fly the drone. Most consumer drones today are *quadcopters*, because they have 4 rotary blades. You can take pictures around your house. You can survey land you own. You can even engage in business activities such as using a camera on a personal drone to take aerial photos of a home that you want to sell. Popular consumer drones include:

- Quadair Drone
- Holy Stone
- Ruko U11
- Snaptain
- Potensic
- 4DV4 Drone
- Lume Cube Drone
- SANROCK U52 Drone

When purchasing a consumer drone, you'll want to think about (besides price):

- Flight time per battery (ranges from 15 minutes to about an hour)
- Speed (somewhere between 45 and 70 mph, typically)
- Smart home return (just press a button on your remote controls, and the drone will return to you)
- Environment awareness ("seeing" obstacles and the ground to avoid crashes)
- Noise (some make little noise, others are very loud)
- Camera (some have built-in camera capability, other have attachments for cameras)

Consumer Drone FAA Regulations

The FAA regulations regarding consumer drones are changing all the time, as we begin to see more widespread use of drones. The FAA's Small UAS Rule (Part 107) addresses the regulations and requirements. As of early 2022, these include: (1) obtaining a Remote Pilot Certificate, (2) registering your drone with the FAA, and (3) meeting several basic requirements such as age (you must 16 or older) and being able to read, speak, write, and understand English. These requirements relate to drones that weigh 8.8 ounces or more. Most consumer drones do.

If you plan on engaging in business activities with your drone, such as taking aerial photos of a home you want to sell, you'll have to take a more rigorous exam for Part 107 certification.

Beyond those initial requirements, you must follow the rules of the air:
- Fly at or below 400 feet
- Keep the drone within sight
- Don't fly in restricted airspaces, near other aircraft, over groups of people, over stadiums or sporting events, or near emergency response areas such as road accidents or forest fires
- Finally, don't fly under the influence

Commercial Drones

Commercial drones are obviously a significant step up from consumer drones, across all characteristics of price, speed, battery life, payload, etc. And the range from "small" to "large" is significant. Smaller commercial drones may have a relatively short battery life (30 minutes) and a payload maximum of a couple of pounds. On the other end of the spectrum are commercial drones used by the military. These can easily weigh tons (literally) and have large payload capacities.

Somewhere in between are the commercial drones that the likes of Amazon, UPS, FedEx, and many other commercial businesses are exploring for package delivery.

▲ ▲ ▲

Delivery Companies and Drones

The race is certainly on to use drones for deliveries. Drone package delivery helps minimize the use of road vehicles, is environmentally friendly because of the use of electricity, and can shorten delivery times because of direct flight capability instead of having to navigate winding roads, traffic congestion, intersections, and the like.

Consider these.
- Zipline is using drones to deliver blood and other medical supplies. [2]
- Kroger and Drone Express have partnered to deliver groceries to consumers. [3]
- UPS has received FAA authorization for a full drone fleet to fly as many delivery drones as it wants. [4]
- FedEx has received authorization as well. [5]
- Amazon has Prime Air, which will deliver packages up to 5 pounds in 30 minutes or less. [6]

Other Innovative Use of Drones

When most of us think of drones, we do so in terms of delivering packages because that's what we most often see in the popular media. Package delivery is going to be big in the drone space, but there are and will be others.
- First responder, police and law enforcement, and public safety [7]
- Agriculture – for reducing time to scout crops, field mapping, and spraying [8]
- Real estate – for surveying land, providing aerial footage of homes, etc. [9]

▲ ▲ ▲

Drones as a Combination of Other 4th Industrial Revolution Technologies

In formulaic terms, it looks like this.

$$D = AI + ST + CT + IoT (+Ba)$$

Like autonomous vehicles, drones are a combination of artificial intelligence, sensing technologies, communications technologies, and IoT. I put batteries (Ba) in parenthetical marks because, while most drones are electric-powered, some really large commercial drones use fossil fuels (gasoline, kerosene, and the like).

Artificial Intelligence in Drones

Today, artificial intelligence may or may not be present in a drone, but it certainly will be in the future. For a drone to be autonomous, it must have sophisticated AI, just as an autonomous vehicle (automobile) must.

We are seeing some "intelligence" added to drones today like environment awareness to avoid collisions with the ground, telephone poles, other drones, and the like. That's most probably a form of reactive AI, which simply detects an object and moves through a series of rules to determine what action to take.

The future challenge of fully autonomous drones is two-fold. The first is the *added-dimension challenge*. Autonomous vehicles work in only two dimensions, while autonomous drones will have to work in three. So, not only must autonomous drones be able to navigate in that third dimension, we'll also have to map the third dimension. Most mapping functions like Google Maps are great with respect to longitude and latitude, but now we'll also need the height of homes, buildings, utility poles, and even trees.

The second is the *circular challenge*. As we add more AI-based capabilities to support fully autonomous drones, we'll essentially be adding more weight to the drone. As the weight of the drone increases, you need more battery capacity. More battery capacity means more weight. More weight means you need more battery capacity. More battery capacity means... I think you get the idea.

Sensing Technologies in Drones

Again, just as autonomous vehicles use a variety of hearing, seeing, and feeling technologies, so do drones, especially as they become more autonomous.

Hearing, seeing, and feeling will be necessary for drones to detect the presence of other objects, both fixed and moving (including birds). Drones will also need seeing technologies like sonar, radar, and Lidar to determine a flight route and respond in real time to changing conditions. Drones will need sophisticated feeling technologies to detect and respond to changes in the wind and wind shear, both horizontally and vertically.

Some drones will most likely use smelling technologies as well, perhaps more so than autonomous vehicles. As a scouter, the National Forest Service uses drones to fly over remote areas to detect the presence of fires. Fire detection can occur through video and image recognition but also through "smelling" the presence of smoke.

And, as we introduce more of these capabilities, we'll have to address the circular challenge. As you add more hearing, seeing, and feeling technologies, you increase the weight of the drone. And we know where that takes us... round and round.

Communications Technologies in Drones

Even the most basic drones require communications, at a minimum from the remote control you operate to the drone itself. More advanced communications technologies will be a necessity as well. If you're delivering a package to someone via drone for example, you'll need GPS capabilities to determine the location of that person.

And eventually, we'll need drone-to-drone communications. Much like autonomous vehicles will communicate with each other to make collective decisions, so will drones.

IoT in Drones

Just about everything we've discussed for hearing, seeing, feeling, smelling, and communications technologies will be in the form of sensors and implemented via IoT. Camera, Lidar, Bluetooth, 5G, and many other sensing and communications technologies will be *a network of Internet-connected objects that collect, process, and exchange data* (our definition of IoT from Chapter 1).

▲ ▲ ▲

DRONES BY THE NUMBERS

Big players in the drone market (especially commercial) include Northrup Grumman, DJI, General Atomics Aeronautical Systems, Thales, 3DR, Boeing, Lockheed Martin, Israel Aerospace Industries, and BAE System. [10]

Flying Cars – eVTOL and Air Taxis

eVTOL (electric vertical take-off and landing) refers to the ability of an electric-powered aircraft to take off, hover, and land vertically. So, eVTOL aircraft do include personal and (most) commercial drones and can also include air taxis, basically the equivalent of a flying car. Consistent with the architecture of a drone, eVTOL aircraft require the use of multiple blades, the *multicopter* concept.

Flying cars may seem a bit far-fetched and almost scifi, but they will be a big part of your future, and they will arrive sooner than most people think. Companies like Toyota, Joby Aviation (which acquired Uber Elevate), Hyundai, Airbus, and Boeing are all racing to perfect the flying car.

Flying cars hold a lot of promise… reduced traffic congestion on the roads, much faster "as the crow flies" times from suburban housing to city-center workplaces (and back), reduction in carbon emissions because of the electric power, faster first responder times to accidents, etc.

It's probably important to keep in mind that flying cars will first appear as taxis or rideshares. The initial cost and maintenance expenses will likely prohibit most people from owning a personal flying car. But, over time, that will change. Won't that be interesting to see.

Flying cars are going to be a fascinating development to watch. Below are a few initiatives:
- City Airbus NexGen by Airbus [11]
- SkyDrive, a Japanese startup backed by Toyota [12]
- Joby Aviation [13]
- Hyundai and Supernal flying taxi [14]
- Boeing [15]

Want to See Something Wild?
Check out AirCar at https://www.aircar.aero/. Fully autonomous, zero emissions, designed to 2 passengers, top speed of 80 mph. It's an eVTOL. And it has 4 engines that operate a total of 8 blades.

IMAGINE FORWARD

Instant Pudding from the Air

It's only a matter of time before our *instant pudding* deliveries come via drone. Everything from fast food to BBQ sauce to sunscreen delivery will find you from the air no matter where you are.

Autonomous Vehicle and Drone Delivery

It will probably be advantageous for the big delivery companies – Amazon and the like – to use delivery trucks combined with drone capability. Both the delivery trucks and the drones will be autonomous. The delivery truck will move through a neighborhood and act as a mobile launch pad for the drones. Each drone will take a package from the truck to a destination and then fly back to the new location of the truck to get the next package. And several drones will be doing simultaneously, really optimizing the time to deliver packages.

Indoor Drones

We're already beginning to see the emergence of indoor drones. For example, the Ring Always Home Cam is an indoor drone with camera capability. [16] Now, you don't need a dozen or so security cameras to see everywhere in your home.

Drone Pooper Scoopers, Leaf Blowers, Window Washers, Pot Hole Finders, House Painters, Gutter Cleaners, Tree Trimmers, and More

This is really going to be a fun space… the innovative uses of drones we'll see for taking over personal, mundane, and un-fun tasks at home. We've already got drone pooper scoopers, what's next? Perhaps drones that can blow the leaves off your yard, wash your windows, paint your house (don't laugh, it will happen), trim your trees, put up and take down decorations, clean the gutters, change light bulbs. Will ladders go the way of the dodo bird?

A Different Way to Watch Sports

This one kind of doesn't need any sort of explanation. Unlimited viewing perspectives of sporting events through the use of drones. Just have to be sure and avoid the ball… baseball, football, soccer ball, tennis ball, golf ball… you get the idea.

Smaller and Smaller

We do now have what are called *nano* or *mini drones*, about the size of the palm of your hand. Many include camera capability. Payload is almost nil, but that will change over time. Imagine drone-mounted speakers for your TV viewing experience. Real spatial audio.

Crowded Airways

Someday I think it will be common to look up to the sky and see perhaps upwards of a hundred or even a thousand drones flying around. It may seem unrealistic, but in the early 1900s people couldn't envision that someday there would be so many cars that we would need multiple lanes of traffic all going in the same direction. (Some people believed there would be so few cars that we wouldn't ever need traffic signs much less traffic signals at an intersection.)

Enter the government. There will undoubtedly be significant legislation and laws regarding use of the airways by drones.

Pads and Parking

Just as public transportation (buses, subways, etc.) has necessitated the need for terminals and parking around the terminals, so to will flying taxis. We'll essentially need helo pads for take-off and landing. And we'll need places for people to park their cars. Of course, then again, maybe not. Perhaps my autonomous vehicle will drive me from my home to the flying taxi pad and drop me off, then return to my home. Hmmm…

People Smarter than Me

The Experts

Well, this is certainly a long list. (Chuckle, chuckle, and chuckle again).

There are lots of great learning resources – people, books, websites – out there for you. And, it really doesn't matter what topics interest you; you can always find learning resources.

I'm listing a few of my favorites here in the hope that you'll explore them at your leisure.

(These are in no particular order, other than random.)

Klaus Schwab

I will forever be grateful to Klaus for his writing on the 4th industrial revolution. Klaus is the Founder and Executive Chairman of the *World Economic Forum*. I recommend 2 of his books:
- *The Fourth Industrial Revolution* [1]
- *Shaping the Fourth Industrial Revolution* [2]

Those books launched my journey into the 4th industrial revolution. (I read both of them while on a cruise of the Caribbean.)

Klaus' books don't deal much in the technical details of the technology. But they provide a wonderful big-picture view of the impacts those technologies can have.

Udemy (www.udemy.com)

Udemy is one of my favorite learning resources. I've taken courses in Udemy ranging from building artificial intelligence to 3D printing to creating augmented reality applications and much, much more.

You can get a free account on Udemy just by signing up. Put in your credit card and start searching for what interests you. When you buy a course, you have it for life.

Word from the Wise: Udemy periodically runs deep-discount sales on its courses. Typically courses are in the $100 range. If you find one you like, add it to your cart and wait for a sale to occur. In doing so, I've been able to buy courses for about $20.

Eric Ries

I love Eric Ries' book, *The Lean Startup*. [3] Eric is one of the principals who launched IMVU, self-advertised as the "#1 Avatar Based Social Experience." I use *The Lean Startup* in teaching our first entrepreneurship class at the University of Denver.

I could go and on about Eric's book but, for our purposes, please consider the definition of a startup...

> **"A startup is a human institution designed to create a product or service under conditions of extreme uncertainty."**

The key is *extreme uncertainty*. The approach in *The Learn Startup* isn't about starting another landscaping business, another retail store with children's clothes, or another software business to help with payroll automation. There is *certainty* in all of those... you know your competitors, you know what customers want, you can estimate costs and build pricing plans, you can easily identify suppliers.

But under conditions of extreme uncertainty, those certainties are not known and can't be discovered through traditional market research and web searches.

Instead, you have to focus on building what Eric calls a minimum viable product (MVP). Build your MVP quickly and inexpensively with only the most basic functionality. Prove that it works. Then, get it into the hands of potential customers for feedback. Use their feedback to build the next iteration of your product, constantly pivoting product features (and your thinking) as you learn more about what your customers want. This is an iterative process that Eric refers to as *build-measure-learn*.

The Learn Startup approach is exactly what you need as you build innovations based on 4th industrial revolution technologies. You'll be building products that the market has never seen before. You'll most likely be building products for a market that – at first – you can't accurately identify (with certainty). Your vision of the features of the product may not match what the market wants.

The Lean Startup… read the book regardless of whether you'll be an intrapreneur or an entrepreneur in the 4th industrial revolution. It's a must-read for intrapreneurs as well. If you take a look back at Eric's definition of a startup, it doesn't focus solely on starting a business, but rather any business (existing or new) that creates products and services under conditions of extreme uncertainty.

Peter Thiel

Peter Thiel was co-founder of PayPal, along with Elon Musk. Peter wrote a great book titled *Zero to One: Notes on Startups or How to Build the Future*. [4] For years, we have used Peter's book as the required reading for our introduction to business course (required of all business majors and minors) at the University of Denver.

Please consider the following excerpt:

> "When we think about the future, we hope for a future of progress. That progress can take one of two forms. Horizontal or extensive progress means copying things that work—going from 1 to n. Horizontal progress is easy to imagine because we already know what it looks like. Vertical or intensive progress means doing new things—going from 0 to 1. Vertical progress is harder to imagine because it requires doing something nobody else has ever done. If you take one typewriter and build 100, you have made horizontal progress. If you have a typewriter and build a word processor, you have made vertical progress."

So, Peter's perspective is similar to Eric's. The "0" is doing what's already being done… opening that landscaping business or the retail store with children's clothing. The "1" is building what doesn't exist. Peter's "1" will be all those new innovations based on 4th industrial revolution technologies.

Zero to One is written as a series of 14 short chapters. Peter designed each to get you thinking about the future and building 1's, not 0's. (There are some really great chapters in the book including *The Mechanics of Mafia* and *You Are Not a Lottery Ticket*.)

Clayton Christensen

Clayton is another must-read author in the innovation space.
I recommend:
- *The Innovator's Dilemma* [5]
- *The Innovator's Solution* [6]

Clayton also wrote innovation-oriented books for the education environment (*Disrupting Class*) and the medical field (*The Innovator's Prescription*). [7,8]

Clayton was the first to apply the s-curve in the invention-innovation space. We looked at the s-curve when we briefly explored the history of the industrial revolutions. Clayton's application of the s-curve was for a specific invention (micro perspective), while my application of the s-curve was for an entire industrial revolution (macro perspective).

Rohit Bhargava

Rohit is the author of the highly-acclaimed *Non Obvious* series, which most recently includes *Non Obvious Mega Trends: How to See What Others Miss and Predict the Future*. [9]

I love Rohit's work because it's about trends that are changing our culture and what the outcome of the changes might be. Most importantly, the focus isn't solely on technology and its impact. It's about all types of trends including everything from attention wealth to ungendering.

Rohit... definitely worth the read.

Et Al

Start with Klaus, Udemy, Eric, Peter, Clayton, and Rohit.

Along the way, also take a look at these.
- Jason Fried and David Heinemeier Hansson, *Rework* [10]
- Gabriel Weinberg and Justin Mares, *Traction* [11]
- Simon Sinek, *Start With Why* [12]
- Tony Hsieh, *Delivering Happiness* [13]

There are undoubtedly many, many others.

CHAPTER 8

Energy Technologies

Harvesting and Storage

Central to the success of the 4th industrial revolution will be our ability to create sustainable energy through energy harvesting and our ability to store that energy, mainly in the form of batteries.

Energy

Energy… interesting phenomenon. If someone asked you to define energy, what would you say?

(Don't worry, most people find it challenging to provide a simple and short definition of energy.)

Technically, **energy** is *the ability to do work*, which makes about as much sense as doing an interpretive dance to the smell of the color nine. (Noodle on that for a while.)

So, let's explore energy, talk about what we're doing in this space, and how energy will play out in the 4th industrial revolution.

The Many Forms of Energy

Energy comes in many forms, including:
- Heat/Thermal
- Light/Radiant
- Motion/Kinetic
- Electrical
- Chemical
- Gravitational
- Magnetic
- Mechanical (many forms including pressure)
- Elastic… just to name the major ones

What's really cool about energy is that it's **not** consumed in the traditional sense of consumption. That is, one type of energy simply transforms into another type of energy.
- When you eat food, the food contains chemical energy which your body stores and eventually converts to kinetic energy when you move.
- When you rub your hands together, you are converting kinetic energy (motion) to thermal energy (heat).
- When you use a flashlight, the energy in the battery is stored in chemical form. The chemical energy is converted into electrical

energy which powers the bulb. The bulb uses the electrical energy to create radiant energy (light).

- When you drop something, the gravitational energy that pulls it to the ground is converted into kinetic energy.

Of course, energy is stored in chemical form in things like coal and natural gas, which we often convert to other energy forms such as electricity.

There you have it. You probably now know more about the basics of energy than 99% of the world's population. (Although, I don't think you need to put that on your LinkedIn profile.)

How We Measure Energy

Each type of energy has its own specific measure, such as watts and kilowatts used to measure electricity. As a single common measure, we convert measurements of specific types of energy to a British Thermal Unit. A **British Thermal Unit**, or **BTU**, is the amount of heat required to raise the temperature of one pound of water by one degree Fahrenheit. (One pound of water is just shy of 16 ounces, or 2 cups of water.) In Figure 8.1, you can see that the U.S. consumed 11.59 quadrillion BTUs (11,590,000,000,000,000) of all types of energy in 2020. [1] (I guess we heat a lot of water.) Most of it, almost 70%, comes from petroleum and natural gas (fossil fuels).

Figure 8.1 **U.S. Energy Consumption, 2020**

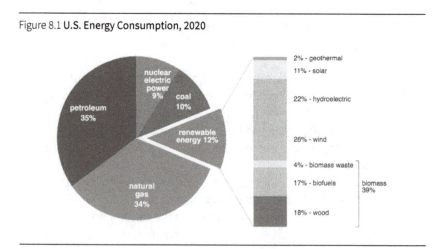

Renewable Energy and Its Importance

Renewable energy is energy that comes from naturally-replenishing and carbon-neutral sources, that include:
- Solar
- Wind
- Hydropower (water movement)
- Geothermal (heat from the earth's core)
- Biomass (wood and wood waste, municipal solid waste, landfill gas and biogas, ethanol, and biodiesel)
- Hydrogen
- Piezoelectricity

Fossil fuels are not considered renewable energy. It can take millions, and sometimes hundreds of millions, of years to slowly convert decomposing plants and animals into fossil fuels.

And as I'm sure you're well aware, we want to get away from our reliance on fossil fuels because our use of those is the largest source of carbon dioxide emissions, a significant component of greenhouse gases.

The Impact of Greenhouse Gases on Global Warming & Climate Change

Greenhouse gas (GHG or *GhG)* is a gas that emits radiant energy causing the greenhouse effect. The *greenhouse effect* is the warming of the earth's surface from radiation in the earth's atmosphere. The primary greenhouse gases are water vapor, carbon dioxide, methane, nitrous oxide, and ozone. Of those, carbon dioxide is the most prevalent in causing global warming.

Global warming is the increase in the overall temperature of the earth's atmosphere, mainly occurring because of the greenhouse effect through the increase in carbon dioxide levels. Global warming and climate change are not the same. *Climate change* refers to long-term shifts in temperatures and weather patterns. So, climate change includes both global warming (an increase in temperature) and global cooling (a decrease in temperature).

As we are in period of global warming, our focus is on the increase in greenhouse gases, how they impact the greenhouse effect, and the subsequent rises in temperature. Greenhouse gases, like carbon dioxide, are actually good; we don't want to get rid of them all. In fact, if we had no greenhouse gases, the earth's temperature would plummet to about 0° F (severe and catastrophic global cooling).

Thus, there is a delicate balance we need to achieve. Greenhouse gases essentially keep the earth warm. But, if greenhouses gases become too much through things like excessive carbon dioxide emissions, then the earth's temperature rises, thus causing global warming.

▲ ▲ ▲

The Personal Cost of Climate Change

The global impact of climate change translates directly into personal costs for you. [2]

- The U.S. Government has spent almost a half-trillion dollars since 2005 on climate change-related disasters. Your tax dollars paid for that.
- Average monthly electric bill: $104.52 in 2009 to $115 in 2019, that's an annual increase of $120 to your electric bill.
- 17.6% increase in beef prices from 2019 to 2021 because of drought, supply chain issues, and labor shortages.
- 2.5% increase in food-at-home prices from 2020 to 2021.
- Catastrophic weather events (wildfires, hurricanes, floods, etc.) were the cause of 39% of insurance claims in 2020.
- In the past 2 years, many insurance companies have increased insurance rates by 10% to 20% to address the impact of climate change.

Climate change is real, very real within the global context and in your wallet.

Energy Harvesting Instead of Energy Manufacturing

Our use of fossil fuels comes from *energy manufacturing*. When you think about energy manufacturing, we often envision large processing plants – smoke stacks reaching to the sky and billowing smoke, the blackness of coal, the iron and steel of vast numbers of railroad tracks and rail cars, and so on. Those types of energy manufacturing plants are using fossil fuels as raw materials, transforming the chemical energy in them to another type of energy.

What we'd really like to do is *energy harvesting*, capturing a renewable energy form and converting into another energy form that is usable "on the spot." So, think of the solar panels you see on your neighbor's roof. They are capturing the energy from the sun (solar source for light or radiant energy) and converting it into electricity for immediate use in the home.

To harvest energy as opposed to manufacturing it, we need to focus more on:
- Solar power
- Wind power
- Hydropower
- Geothermal power
- Biomass
- Hydrogen
- Piezoelectricity

Solar Power

Time for another question: One second of the sun's total energy output would power the United States for how many years?
- A. 900 years
- B. 9,000 years
- C. 90,0000 years
- D. 900,000 years
- E. 9,000,000 years

The answer is 9,000,000 years. [3] Unbelievable when you think about it. If we could figure out a cost-effective and environmentally-friendly way to harness the power of the sun, we would solve all our energy problems. End of story.

Formally, **solar power** is the conversion of sunlight (radiant energy) into electricity. We hope and believe that solar power will become the world's largest source of electricity by 2050. (We should work diligently to make it happen sooner.)

A lot of progress is being made in this area, and there are many ways to capture the radiant energy of the sun, with the two most common being:

1. **Photovoltaic system (PV system)** uses solar panels to absorb and convert sunlight into electricity. Those solar panels on your neighbor's roof are a PV system.
2. **Concentrated solar power (CSP)** uses mirrors and lens to concentrate a large area of sunlight onto a receiver, which converts it to heat which drives a heat-engine/steam turbine connected to an electrical power generator. CSPs are much larger in size than PV systems, so you won't find them on your neighbor's roof. These often occupy several acres of land.

Besides converting solar energy into electricity, we're also seeing a number of applications using the thermal energy of the sun.

- Solar water heating, especially in the middle geographical latitudes (Ethiopia, as an example. [4])
- Heating, cooling, and ventilation in the home
- Desalination (making potable water from saline and brackish water) [5]
- Solar water disinfection (SODIS), exposing plastic bottles of water to sunlight. Millions of people in developing countries use this method for their daily drinking water. [6]

Mobile Solar (MoSo) – The Time Is Now

Our traditional view of solar power is large fixed-in-place solar panels generating energy for homes, cars, commercial buildings, street lights, etc. The next evolution is for all of us to take advantage of mobile solar. **Mobile solar (MoSo)** is the use of small portable solar power units that you can take with you wherever you go.

MoSo units usually weigh a couple of pounds or less and enable you to recharge smaller electronics such as your phone and tablet. Throw them on your backpack while hiking, place them on a table while you're enjoying a day in the sun… whatever. MoSo units will save you a couple of dollars per year in electricity charges per device. That seems small. But if we all do it, we could collectively save billions of dollars per year in electricity costs.

Check out these options:
- Hiluckey Outdoor Portable Power Bank
- CONXWAN Solar Charger
- FOCHEW Solar Portable Charger Power Bank
- YPWA Portable Charger Power Bank
- Sunnybag Leaf Mini

Each costs less than $50.

Wind Power

We've taken advantage of wind power for thousands of years. Sailing ships from the Viking age, for example, captured the power of wind in their sails to propel the ship across the water (kinetic energy). Of course, now we're focusing on capturing wind power and converting it into electrical energy.

Wind power is the capturing of the kinetic energy of wind to power wind turbines which provide mechanical power to electric generators which in turn generate electricity. Most often, we capture wind power using a **wind farm**, a collection of wind turbines in close proximity that power numerous electric generators that are connected to an electric power transmission network. The power transmission network moves the electricity from the energy generation source (the wind farm) to homes, businesses, cities, and grid storage units.

The two dominant types of wind farms are onshore and offshore. Although I shouldn't have to say this, I will… onshore wind farms are a collection of turbines on land, mostly rural areas, while offshore wind farms have their turbines on water. Onshore wind farms are much less costly to build and maintain than offshore wind farms, but many people believe that onshore wind farms are leading to habitat loss (animals) and the industrialization of the countryside. Offshore wind farms, again while expensive to build and maintain, do have a steadier and stronger supply of wind than onshore wind farms.

Like solar, if we could figure out a cost-effective and environmentally-friendly way to harness the power of the wind, we would solve all our energy problems. (Different ending to the same story.)

Hydropower

Hydropower or *water power* is the capturing of the kinetic or gravitational energy of running or falling water to run a turbine or series of turbines which in turn generate electricity. This is achieved in a hydroelectric power plant. Hydroelectric power plants are often thought of as human-made dams that create a reservoir of water so that there is a consistent flow of water to the turbines. But, they can also be what are called run-of-river, which doesn't use a dam to create a reservoir. Run-of-river hydropower electric plants rely on a consistent flow of water to meet energy needs.

The most well-known hydroelectric power plant in the U.S. is the Hoover Dam, completed in the late 1930s between the borders of Nevada and Arizona on the Colorado River. (Put the Hoover Dam on your bucket list. Take the long tour with a guide. Fascinating story.)

Geothermal

Geothermal power takes on two forms, either (1) electricity generated from geothermal energy or (2) heat generated from geothermal energy. In either case, the naturally-occurring thermal energy in the earth's crust is used.

Geothermal heat generation has actually been around for thousands of years, dating way back to the Paleolithic times, and includes:
- Oldest known spa – 3rd century BC Qin Dinasty in China
- Public baths and underfloor heating – Bath, Somerset, England (1st century)
- Oldest geothermal district heating system – France, 15th century
- First known building to use geothermal energy for heating – Hot Lake Hotel in Union County, Oregon

In the early 21st century, we built the first successful geothermal power generator which used geothermal energy to generate electricity.

Biomass

According to 2020 data in Figure 8.1, biomass accounted for 39% of all renewable energy use in the U.S.. Renewable energy accounted for 12% of total U.S. energy use, so biomass accounted for roughly 5% of U.S. energy consumption. *Biomass* is the use of plant and animal material as fuel to generate electricity or heat. Biomass includes things like:
- Biomass waste – waste from forests, yards, and farms (bark, sawdust, wood chips, wood scrap, etc.).
- Biofuels – energy crops or crops grown specifically for energy production (willow trees, poplar trees, and elephant grass are examples); ethanol and biodiesel also fall into this category.
- Wood – you guessed it, the burning of wood for cooking, lighting, and heating. Wood was the primary source of energy in the U.S. up through mid 19th century.

Hydrogen

Not to be confused with hydropower, hydrogen fuel also holds potential as a renewable energy source. A **hydrogen fuel cell** produces electricity by combining hydrogen and oxygen atoms. A hydrogen fuel cell is 2 to 3 times more efficient than a traditional internal combustion engine running on gasoline. Even better, the only byproduct is water, so it falls into the category of *zero-emission*.

While there is much interest and research in hydrogen fuel cells, there is very limited commercialization in the electric vehicle space. As of early 2021, there were fewer than 50 hydrogen vehicle fueling stations in the U.S., with most of those being in California. [7]

Still, an energy source and technology to watch.

> (BTW, Elon Musk referred to hydrogen fuel cells as "mind-bogglingly stupid." But, a 2017 survey of 1,000 automobile executives yielded that hydrogen fuel cells ultimately would outperform battery-powered electric vehicles. [8])

Piezoelectricity

I had to include piezoelectricity because I think it has a lot of potential, and it's really, really cool. **Piezoelectricity** is created when you apply pressure to specific types of materials (for example, certain crystals and ceramics). You can, for example, embed piezoelectric energy harvesting plates in a sidewalk. When you walk on the plates, the pressure of your steps creates electrical energy that can be harvested.

Applications of using piezoelectricity include:
- Tokyo subway – as travelers walk on the plates, the captured electricity runs turnstiles and displays
- Sainsbury's – an English supermarket capturing the electricity generated by shoppers
- Dance club – In Rotterdam, uses captured electricity to run the dance floor light shows
- Sidewalk – in Toulouse, France pedestrians create electricity that is used by the city to power light poles [9]

Some people are even proposing that we embed piezoelectric energy harvesting plates in roadways to generate electricity from the cars driving over them.

▲ ▲ ▲

The Paris Agreement and Climate Change

The *Paris Agreement* (also called *Paris Accords*, *Paris Climate Agreement*, or *Paris Climate Accords*) is an international treaty on climate change adopted in 2015 by 197 countries. The main goal of the Paris Agreement is to substantially reduce greenhouse gas emissions in an effort to limit global temperature increases in the 21st century to no more than 2 degrees Celsius above preindustrial levels. The big hairy audacious goal is to limit global temperature increases to 1.5 degrees Celsius.

The Paris Agreement is a commitment by all countries to reduce pollution, with a big focus on carbon dioxide reduction. It also includes a commitment by developed nations to assist less developed countries in their efforts to address climate change, through financial resources and the sharing of technologies. Finally, the Paris Agreement provides for a transparent framework for monitoring and reporting the efforts of each country to address climate change.

Batteries: Energy Storage

The use of energy storage technologies, with our current paradigm being batteries, is essential to the success of the 4th industrial revolution.

- All the tech – IoT sensors, AI, blockchain, etc. – must have a source of power.
- As we begin to rely less on energy manufacturing and traditional utility grids and focus more on energy harvesting, we'll need to have better storage of just-in-time energy harvesting for use at a later time.

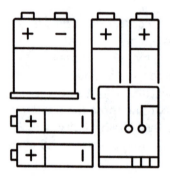 As a formal definition, a **battery** is a source of electric power for powering electrical devices such as flashlights, computers, and automobiles. In the case of energy storage, batteries give us the ability to balance demand (energy when we need it) and supply (energy when we can capture it).

Consider solar power. Today it's bright and sunny and your PV systems capture a lot of energy, so much energy that you can't use it in a normal day of operating your house. You need a way of storing that energy, because tomorrow may be cloudy and rainy.

For that reason, we need to think about the intermittent supply associated with energy harvesting. Wind doesn't happen daily. For generating energy, sometimes we get too much, sometimes we get too little. The same is true for sunlight. We have to use the sunny days to save for the rainy days.

And of course, given that so much tech is going to be everywhere in the world, we cannot rely on the fact that the tech will be close enough to "plug into a wall outlet."

Types of Batteries

Energy inside a battery is stored as chemical energy (reactants, electrons, positive/negative electrodes, and all that ~~cool~~ stuff you learned in your high school chemistry class).

According to the chemical structure used inside a battery to store energy, you get one of two main types of batteries:

1. ***Primary (non-rechargeable) battery*** – a battery whose chemical structure is not designed to be returned to its original charged state. Thus, the battery is considered a one-time-use battery. Batteries in watches, key fobs, hearing aids, etc. are primary batteries. (A lot of these are commonly referred to as ***button batteries***). An alkaline battery, the most popular type of battery in the world, is technically a primary battery. (See the discussion on alkaline batteries below for rechargeables.) Alkaline batteries are used in children's toys, flashlights, TV remote controls, etc.

2. ***Secondary (rechargeable) battery*** – a battery whose chemical structure is designed to be returned to its original charge state by – you guessed it – recharging it. A lithium-ion battery is the most common secondary battery you hear about. Lithium-ion batteries are used extensively in tablet computers, laptops, electric vehicles, etc.

Alkaline Batteries

Alkaline batteries account for about 80% of all batteries worldwide. Alkaline batteries store energy in chemical form as a reaction between zinc metal and manganese dioxide. (I bet you're glad you know that.) They come in a variety of types and sizes including A, AA, AAA, C, D, and many other alphabetic configurations.

Can you recharge an alkaline battery? The short answer is yes, the longer answer is, well, longer. Some alkaline batteries are designed, manufactured, and advertised as rechargeable alkaline batteries. They are often green or light blue in color such as those offered by ULINE, Pale Blue, and Brightown. Of course, you have to buy a special recharging unit for these.

The trend is toward rechargeable alkaline batteries. IKEA, among others, no longer sell non-rechargeable batteries in their retail stores.

Experts agree that you shouldn't recharge an alkaline battery that isn't advertised as <u>rechargeable</u> for a variety of reasons. We'd have to get out the old weed whacker to delve into that debate, so let's leave it alone.

You can also recycle alkaline batteries. Many U.S. states are beginning to require the recycling of alkaline batteries, as opposed to throwing them in a landfill.

Lithium-Ion Batteries

A ***lithium-ion battery*** is a rechargeable (secondary) battery in which lithium ions move from a negative electrode to a positive electrode when you extract the energy. When you recharge a lithium-ion battery, the lithium ions move in reverse from the positive electrode to the negative electrode, and thus the battery is returned to its original fully-charged state. (Again, I bet you're really glad you now know that.)

Lithium-ion batteries are all the rage in portable electronics (smartphones, laptops, tablets, etc.) and especially electric and hybrid electric vehicles (called, obviously enough, an ***electric-vehicle battery, EVB***). They are also popular in battery-powered power tools such as drills, skill saws, weed whackers (there it is again), lawn mowers, etc.

You also see lithium-ion batteries used as a primary storage system for wind energy and PV solar systems. When your PV solar system captures the sun's energy and converts it into electricity, that electricity is most often stored in a lithium-ion battery (more appropriately, a lithium-ion battery bank).

Your Car Will Be the Emergency Power Backup for Your Home

Lose power for your home, no need to worry. Just plug your home into your car. Use the power in your electric-vehicle battery to run your home until the power is restored.

Solid-State Batteries

It's pretty simple. We need batteries that are smaller, hold more energy, have shorter recharging times, have longer life spans, are environmentally friendly when thrown away, and are not susceptible to leaks, explosions, and the like.

One such promising development is called a solid-state battery. A **solid-state battery** utilizes solid electrodes and electrolytes instead of liquid and polymer gel electrodes and electrolytes like those used in current lithium-ion batteries.

While right now expensive to make and still in their development, solid-state batteries hold the promises of:
- Higher energy densities (storing more energy in a much smaller space)
- Lower risk of catching fire, because material is solid-sate and not liquid or polymer gel
- Faster recharging
- Higher voltage
- Longer life

No doubt, batteries of the future will look and be much different from those of today.

Hot Companies to Watch in Battery Storage

Of course, Tesla will come to mind for most of you. Tesla, to be sure, is a leader in industry disruption not only in the automobile industry but also renewable energy. Tesla's Powerwall, for example, captures solar energy using photovoltaic (PV) systems and stores in it a "wall" or bank of lithium-ion batteries. You can use the wall of energy to power your house, recharge your electric vehicle, and as a backup power source.

Tesla, LG Chem, and Enphase Energy are the big three in the battery energy storage market. Together, they account for 85% of all sales in the battery energy storage market. [10]

But, this is becoming a very competitive space, with lots of new startups emerging all the time. Some of those include SK Innovation, QuantumSpace, Ionic Materials, and NEI Corp. [11]

CHAPTER 9

Sensing Technologies

Hearing, Seeing, Feeling, and Smelling

How many senses do we have?
- A. 3
- B. 4
- C. 5
- D. 6
- E. 7

Another trick question, of sorts. Most people think we have 5 senses, but we actually have 7.

The 5 common ones are – of course – sight, sound, touch, smell, and taste. The two lesser-known senses are:
- *Vestibular* – (movement) the movement and balance sense, which gives us information about where our head and body are in space. Vestibular sensing helps us stay upright when we sit, stand, and walk. (Gymnasts have a keen vestibular sense.)
- *Proprioception* – (body position) the body awareness sense, which tells us where our body parts are relative to each other. Proprioception sensing helps us understand how much force to use, allowing us to do something like crack an egg while not crushing the egg in our hands.

Many hearing, seeing, feeling, and smelling technologies have been around for several years in the 3rd industrial revolution. I'm not presenting them here specifically as technologies of the 4th industrial revolution. But, we will need them for 4th industrial revolution innovations.

Hearing Technologies

Hearing technologies are those technologies that detect and interpret sound, which could be the sound of a door closing, your voice, a simple thump, or even the sound termites make while eating the frame of your house.

IoT Sound Sensors

These are microphone-based sensors that detect the presence of sound or noise and usually cause some sort of response. Sound sensors are popular for applications like home security and monitoring, for example, turning on an outside light when sound is detected. They are also popular for basic control of home appliances like lamps (clap once for one, clap twice for off).

They differ from more advanced speech recognition which not only detect sound or noise but also must capture sound at a certain quality to make sense of words.

For IoT sound sensors, you can choose from among a wide variety of microphone types like dynamic, carbon, ribbon, and condenser. You'll need to select the microphone type based on the situation and what you're trying to capture. (I'll let you research more on different types of microphones at your leisure.) You can also choose from characteristics such as big or small (size). And you can control the sensitivity of an IoT sound sensor. IoT sound sensors that you program to be less sensitive will only detect louder sounds.

Basic IoT sound sensors are about $1 in price.

Speech Recognition

Speech recognition, often called *automatic speech recognition* or *ASR*, are technologies that detect, interpret, and respond to your spoken words. Obviously, things like Amazon Alexa, Google Assistant, and Siri rely on speech recognition technologies, as does your remote control so you can say things like "Sopranos" instead of using that long arduous search process of moving the cursor around on a keypad on your TV, computer, or tablet.

Many such hearing technologies incorporate machine learning. The more you speak to a hearing technology, the more it learns to recognize not only your voice but also your tone, inflection, and use of certain words.

Speech recognition is also vitally important as an assistive technology for people with sight or movement limitations.

If you want speech recognition for your personal computing needs, you'll find several great ones in the cloud. (You might want to do a little reading on cloud computing in the Infrastructure Technologies Chapter.) These include: [1]
- Google Cloud Speech API
- Microsoft Azure Cognitive Services
- Amazon Transcribe
- IBM Watson
- SpeechMatics

As with all cloud or as-a-service offerings, these usually charge by the word or minute.

Variations of speech recognition are present in many aspects of your life. Some elevators respond to verbal commands… open door, 3rd floor, etc. Music recognition apps like Shazam, SoundHound, and Musixmatch all use a variation of speech recognition.

▲ ▲ ▲

Seeing Technologies

The formal term for seeing technologies is computer vision. In short form, **computer vision** is all about image recognition and understanding (by a computer). Computer vision is our attempt to replicate the ability of the human visual system. So, computer vision includes such things as the basic capturing of image and video all the way up through using artificial intelligence to determine what an object is, how far away it is, how fast it's moving, and so on.

> (Seeing technologies also include the "production" of things for you to see. These include the simple and straightforward like your computer screen and also the more advanced like headsets and visual displays for augmented, mixed, and virtual reality environments.)

Lots of important technologies in the "seeing" realm. Just a few are below, all of which are vitally important to the success of the 7 core 4th industrial revolution technologies.

360 Camera

A **360 camera**, also called an **omnidirectional camera**, has a 360-degree seeing ability, thus giving it the ability to capture both still photos and videos of the entire surrounding area. 360 cameras are popular for applications such as:

- Surveillance and security – in-home security cameras, for example, that can also detect motion.
- Google Street View – uses 360 cameras to give you the ability to see in different directions while in Street View.
- Virtual reality – 360 cameras are popular in this venue because of your need to move your head and see in different directions. 360 cameras used for this purpose are often called **VR cameras**, special types of cameras that render a 3-dimensional view.
- Live event (virtual attendance) – not much need to explain this.
- Real estate – for enabling potential buyers to virtually tour a home, building, etc.

Sonar, Radar, and Lidar

These all work on the same basic principle; they measure the time it takes for a signal to reach an object, bounce off of that object, and return to the sending unit. They differ in that sonar uses sound, radar uses radio waves, and Lidar uses light waves.

- *Sonar*: **So**und **N**avigation **a**nd **R**anging
- *Radar*: **Ra**dio **D**etection **a**nd **R**anging
- *Lidar*: **Li**ght **D**etection **a**nd **R**anging

Each has their advantages and disadvantages, making them more suitable in certain situations and not in others.

For a long time now, sonar has been the dominant "seeing" technology in the water. Sonar is used by navies to detect vessels (both submarines underwater and top-water vessels). Sonar is also used to map the ocean floor and search for underwater objects. When you read about sonar technologies, you'll see a couple of other terms pop up… ultrasound and ultrasonic. These are subsets of sonar technology. For example, ultrasound is used in the medical field to "peek" inside the human body.

Radar is popular in other realms including spotting aircraft in the sky (air traffic controllers) and determining the speed of moving vehicles (law enforcement).

Lidar is all the talk for autonomous vehicles. Lidar emits millions of laser light pulses to create a 3D map of the surrounding area. It has an extremely high degree of accuracy.

(See Chapter 6 on Autonomous Vehicles for more on the use of sonar, radar, and Lidar.)

With the release of the iPhone 13, Apple began putting Lidar capabilities in its phones. (The other competitors in the phone space quickly followed.) With Lidar on your phone, you can, for example, map the exact layout of a room. You can capture height, length, width, depth, distance, and so on.

It's going to be fascinating to see the innovations within phone-based Lidar. For example, you could map a room and then add falling snow.

The snow would accumulate in various ways on the furniture and floor but not the floor underneath the coffee table. Why? Because the Lidar mapping would note surface area above the floor which would prohibit snow from falling there.

OCR

Optical character recognition (OCR) is the electronic conversion of typed, handwritten, or printed text into a computer-usable format. Optical character recognition had been around for quite some time, dating back into the 3rd industrial revolution, but remains an important technology for the 4th industrial revolution.

Applications and uses of OCR include:
- Assistive technologies for the sight-impaired (e.g., converting printed text into audio format)
- At-register scanning and processing of checks
- Photo-based depositing of checks
- Traffic sign recognition (important for autonomous vehicles)
- License plate recognition (automated ticketing systems for driving violations and also toll roads)
- Scanning printed documents to create searchable PDFs
- Handwriting recognition (accompanied by machine learning techniques)

SLAM

SLAM (simultaneous localization and mapping) is a seeing technology for environments that need the constructing and updating of an environment including the location of the "agent" within it. Here, think about the popular iRobot Roomba, the self-driving robot room vacuum cleaner. It uses SLAM to map a room, detect objects to avoid, and keep track of where it is in the room. The "agent" is the iRobot Roomba.

Biometrics

The use of biometrics is gaining more societal acceptance daily. **Biometrics** literally means the measures of life, a combination of "bio" or *life* and "metrics" or *measures of*. Biometrics can be something as straightforward as taking your blood pressure or advanced applications like using your fingerprint for identification and really advanced applications in biomedical engineering, bio-technology, and neuro-technology. Biometrics also touches 4th industrial applications like using 3D printing to create synthetic skin grafts (and someday 3D-printed fully-artificial organs.)

As we begin to see more of the integration of technology and the human body, biometrics will become increasingly important.

Motion Capture

Motion capture (also *mocap* or *mo-cap*) is the process of detecting and recording the movement of objects or people.

Motion capture is used for a variety of applications. In filmmaking and video game development, motion capture is used to capture the movement of actors to create 2D or 3D animated renderings. In the field of physical therapy, clinics use motion capture to determine a patient's progress. In sports safety, researchers use motion capture to understand, for example, how the body changes from impact to design better safety and sports equipment such as helmets, safety harnesses, and the like.

Subsets of motion capture include gesture recognition and hand/finger tracking.

Gesture Recognition

Gesture recognition is a technology with the goal of recognizing human gestures. Human gestures can include the shrugging of shoulders, changes in facial patterns due to emotional changes, in-the-air swiping by the hand for touchless interfaces, body movements while using virtual weapons while in a virtual reality game, and so on.

An important part of gesture recognition is hand and finger tracking.

Hand and Finger Tracking

Hand and finger tracking is a type of motion capture for detecting and understanding finger and hand movements.

One of the first applications of hand and finger tracking was in the interpretation of sign language. Evalk, for example, offers GnoSys, a smartphone app that can translate sign language into speech. [2]

Hand and finger tracking take place in 2 environments. The first is like GnoSys, which doesn't require the use of gloves for capturing hand and finger movement. The second is the use of gloves to track hand and finger movements, which fall into the category of feeling or haptic technologies.

Eye Tracking

Yet another part of gesture recognition is **eye tracking**, the process of measuring the point of gaze, i.e., where someone is looking. Eye tracking technologies are particularly helpful in human-to-computer interaction (HCI) for people with movement limitations that prohibit them from using traditional input devices like keyboards, mice, and trackpads.

Eye-tracking technologies are becoming increasingly prominent in marketing and customer experience management. Eye-tracking technologies enable marketers to detect what captures a consumer's attention, in what order a consumer reads information on a box, and so on.

▲ ▲ ▲

Feeling Technologies

The formal term for feeling technologies is haptic technology. ***Haptic technologies*** are those technologies that (1) create the sense of touch by applying force and motion (including vibration) to a person, or (2) detect and interpret force or motion from a person. So, feeling technologies are a two-way street. The computer is sending you "feeling" in the form of haptic interfaces and/or you are sending the computer "feeling" in the form of your motions, actions, etc.

When I "Feel" the Computer

There is a big range of devices in this area. You can have a game controller that vibrates to give you a sense of some sort of explosion. You can also use a specialized gaming chair that also provides vibration.

Many automobiles are incorporating these types of technologies as well. For example, when you start to change lanes and your car notes that another car is in the other lane, your car may vibrate the steering wheel or even make it somewhat difficult for you to turn the steering wheel in that direction.

When the Computer "Feels" Me

The easy (and dull) ones include things like your using a mouse or touchpad to move the cursor on the screen, clicking on buttons, highlighting, and the like and also moving a joystick and selecting buttons on a controller. It can also include:

- Hand controllers in extended reality environments for basic interface functions and also specialized functions like using a bow and arrow.
- Headsets – electronic headsets that adjust their field of vision to your moving your head in different directions, up and down and side to side. Headsets are actually a combination of feeling (capturing and interpreting your head movements) and seeing (changing your field of vision) technologies in virtual reality and mixed reality applications. (See Chapter 4 on Extended Reality for more on virtual and mixed reality applications.)

When the Computer "Feels" Other Things

Feeling technologies are not limited to computer-to-person interaction. You can apply feeling technologies in a wide variety of environments. These would include things like:

- Vibration/Motion sensors – placed on equipment to determine when a machine may be coming out of balance.
- Pressure sensors – to determine weight change, for example
- Tilt sensors – to determine the angular orientation of something (e.g., is a platform for a wind turbine starting to shift?). Tilt sensors support the vestibular sense.

An interesting application in this space is that of a digital twin. A ***digital twin*** is a real-time digital counterpart of a physical object. Take the wind turbine as an example. To a wind turbine you can add IoT sensors all over it (and around it) to create a real-time digital version of it. The IoT sensors can measure stress, motion/vibration, speed of the wind, stability of the platform, changes in soil structure around the turbine, and so on. This enables you to monitor the wind turbine without being at the wind turbine location to constantly monitor it physically.

▲ ▲ ▲

Smelling Technologies

Smelling technologies (also known as ***digital scent technologies*** or ***olfactory technologies***) are technologies that can detect natural scents/smells/odors/molecules/particles in the air and also produce synthetic scents/smells/odors/molecules/particles. Obviously, smelling technologies attempt to replicate the human olfactory system, the detection part of it anyway.

I know, I know… scents, smells, odors, molecules, and particles… long list, overlapping terms, rather haphazard. Bear with me while I explain.

Firstly, without getting too much into the weeds, your sense of smell detects the presence of odorous molecules in the air. The binding of odorous molecules with your olfactory receptor neurons creates a chemical stimulus causing electrical signals to be sent to your brain. (Enough said.)

It is this sense of smell that allows you to smell when food is going bad, perfume, dirty socks, flowers, and someone's bad breath. These are about odors, scents, and smells, which result from a change in the molecular structure of the air around you. These are the types of odors (i.e., molecular structural changes) that you can "smell."

Secondly, other types of molecular structures cannot be detected as "smells." The most common of these in our homes is carbon monoxide. Carbon monoxide is a colorless, odorless, and tasteless gas. That's why you have carbon monoxide detectors in your home; you can't see or smell carbon monoxide, so your carbon monoxide detector recognizes their molecular presence in the home. Natural gas is another gas that has no odor. Companies that sell natural gas add a harmless gas called mercaptan to natural gas to give it that rotten egg smell.

Finally, other types of smelling technologies use the presence of particles in the air to detect "smell." The most common of these is a smoke detector in your home. Smoke produces particles in the air, which smoke detectors can detect or "smell."

In this smelling space, we have lots of IoT sensors. There are IoT sensors specifically for smelling carbon monoxide, carbon dioxide, and natural gas. You can program a general smelling sensor for things like food going bad or the presence of some diseases, infections, abscesses, or tumors that have an external manifestation (i.e., not internal). And, of course, there are smoke detector sensors which detect smoke-related particles in the air.

That's the detection presence side. There are also smelling technologies that produce synthetic smells. You should check out OVR Technology, which has created an Architecture of Scent Platform. Developers can use this platform in various types of extended reality applications to create a more immersive experience through the sense of smell. Visit https://ovrtechnology.com/ and https://www.youtube.com/watch?v=bHL7hEvNv7U. Absolutely fascinating.

▲ ▲ ▲

Barcodes (UPC and QR Codes)

Barcode (also **bar code***)* is a method of representing data in a machine-readable, standardized format. You've seen these all over the place, mostly as either a UPC or QR code.

A **universal product code (UPC)** is a type of barcode that uses vertical bars, with the size of the bars and the distance between them determining the number. These are referred to as linear or 1D (1-dimensional) barcodes. A UPC has 12 numbers, with the most common application being a product number in an inventory system. Think of just about any product in a grocery store. It contains a UPC that uniquely identifies the product, such as a box of Cheerios. Of course, the limitation is that every box of Cheerios has the same UPC, so the store can't distinguish between each box.

A **QR (quick response) code** is a matrix barcode and referred to as a 2D (2-dimensional barcode). QR codes can contain much more information than a UPC. Many QR codes commonly contain date and time information.

Organizations can and do create their own unique QR codes, often for one-time use. You've undoubtedly received a QR code for event tickets, promotions, etc. These were generated by the organization (e.g., SeatGeek, TicketMaster, or StubHub) and literally discarded after the specified use. Of course QR codes can be more permanent. During the COVID pandemic, restaurants stopped distributing menus. Instead, patrons scanned a QR code at their table to view a copy of the menu online.

There are many different and interesting configurations of QR codes including Aztec Code, SnapTag, and SPARQCode. Explore them on your own and "nerd out" if you want.

CHAPTER 10

Communications Technologies

Anywhere Is the New *Where*

Faster, better, and more reliable communications technologies will enable more widespread adoption and use of many of the key core 4th industrial revolution technologies including autonomous vehicles, drones, IoT, and extended reality.

Many communications technologies have been around for several years in the 3rd industrial revolution. I'm not presenting them here specifically as technologies of the 4th industrial revolution. But, we will need them for 4th industrial revolution innovations.

Bluetooth

Bluetooth has been around for quite some time, dating back to its first commercial uses in the early 2000s. **Bluetooth** is a short-range wireless technology that has been standardized for communications between devices (both fixed and mobile) over short distances, typically up to about 30 feet. For you the most common applications include connecting your headphones/earbuds to your phone, connecting your phone to your computer, and connecting your phone to your car's entertainment and navigation systems.

Bluetooth has many variations. The 2 main ones are classic Bluetooth and Bluetooth LE, with the LE standing for *low energy*. Classic Bluetooth is widely used in situations for which you need a constant connection and move a lot of information. Wireless speakers would be an example. Bluetooth LE is for situations when you only need an intermittent connection and not move a lot of information. Tile and all its competitors, the popular solutions to *where's-my-fill-in-the-blank*, use Bluetooth LE. Many IoT applications use Bluetooth LE. Do you really need an always-on and continuously communicating conduit? Regardless, If your innovation requires a Bluetooth type connection, you'll want to learn more about the different types of Bluetooth.

Bluetooth is essential to the success of the 4th industrial revolution, especially in the IoT space. Bluetooth is great for connecting IoT sensors to each other and to a "unit" that manages the IoT sensors and collects and processes the information from the sensors. Consider, for example, sensors buried in my yard that communicate with a water sprinkler control system so it can adjust watering frequency and lengths of time based on the dryness of the soil in various parts of the yard.

Radio-Frequency Identification and Near-Field Communication

The wireless connectivity in your credit or debit card that allows you to make contactless payments most likely uses NFC, or near-field communication, which is a subset of RFID. So, let's talk about RFID first and then NFC.

Radio-frequency identification (RFID) is a communications technology that uses electromagnetic radio waves (sent by an RFID reader device) to identify and track tags attached to objects. The process is quite simple and widely used. The reader device sends out a radio wave and the RFID chip responds by sending back digital data, which at a minimum is some sort of identification, to the device reader. This identification can be a unique tag number (which would be used to access a database of information about the object), an ISBN (for a book), badge number, an inventory number, etc. Some RFID tags transmit other information such as production lot number, date of production, and so on.

 RFID tags are either passive or active. Passive RFID tags have no battery, but instead use the energy from the electromagnetic radio waves to create enough electrical energy to cause the RFID chip and transponder to send out the digital data. Passive RFID tags work well for short distances (usually a few feet) and are inexpensive in bulk, often as low as $.05.

Active RFID tags have a built-in power supply and work over greater distances and can hold and communicate more information. Active RFID tags can work over distances of several hundred meters. They can also cost upwards of $25 or more.

RFID tags can also be read-only or read-write. A read-only RFID tag can only transmit its digital data. You cannot send new information to a read-only RFID tag and have it update the stored information. Of course, read-write RFID tags have both the ability to send digital data and to receive digital data and update the information stored on the tag.

RFID tags are widely used.
- Libraries – to check-in and check-out books
- Livestock tracking
- Passports
- Typical product inventory management (retail shelves, warehouse applications, etc.)
- Human implants (yes, this is happening)
- Access control (badge identification)
- Healthcare
- Transportation (toll roads, for example)

Near-field communication (NFC) is a subset of RFID but has data transmission speed and distance limitations. Most all NFC applications are limited to a few centimeters, usually no more than 10.

As I stated before, your credit or debit card probably uses NFC. The technical term for that particular payment communications standard is **EMV**, which stands for **E**uropay, **M**astercard, and **V**isa. Those companies worked together to develop EMV and were the first to roll it out in the payments space.

The concept is the same for NFC as for RFID. When you "tap" the reader with your credit card, your credit card is close enough to the reader device that it activates your EMV which in turn transmits your credit card information to the reader device. (BTW, the physical "tapping" is of no importance. Contact not required.)

WiFi

WiFi (wireless fidelity) is a suite of wireless network protocols that support local area networking for devices and Internet access. (We most commonly use the term WiFi, but its technical term is Wi-Fi. I guess we got tired of typing that silly dash all the time).

Not much to say here, kind of like Bluetooth. WiFi is the most commonly used network infrastructure in the world, especially for homes, offices, coffee shops, airports, etc. Network range is typically about 75 feet but can range of up 500 feet, especially in some outdoor applications.

LiFi

(Finally, something new and interesting).

WiFi will continue to get faster, better, and more reliable. And I think we'll be using WiFi for the foreseeable future, but LiFi is on the horizon. **LiFi (**also **light fidelity** or **Li-Fi)** is a wireless communications technology that utilizes light to transmit data, as opposed to WiFi that uses radio frequency waves. LiFi can transmit data at extremely high speeds using visible light, ultraviolet, and infrared.

The promise of LiFi is substantial. Firstly, it can transmit data in excess of 100 gigabits per second. (As of 2022 when I wrote this book, the average speed of a home WiFI unit was less than 1 gigabit per second, and more typically around 100 to 200 megabits per second.) Secondly, because Lifi uses light transmissions and not radio frequency waves, it would not interfere with other RF-based networks. Think about being on an airplane. A LiFi network wouldn't interfere with the communications network of the cockpit, and you could literally turn on your overhead light to boost your speed.

Someday, you may pick out a ceiling fan for your home and choose your lights based on how much you want to boost your network speed. So cool.

Currently in the invention stage, but watch the development closely.

The Gs, 2 through 5

The G in 2G, 3G, 4G, and 5G refers to **g**enerations in cellular or mobile communications. (Both 2G and 3G are now history.) Let me cover a couple of terms quickly and then we'll compare 2G, 3G, 4G, and 5G.

- **Cellular network** – communications network that wirelessly connects land areas called "cells," with each cell having at least one transceiver. A transceiver is a device that can transmit and receive data to and from wireless devices and to and from other transceivers. When you're out and about using your phone, you're doing so using a cellular network.
- **Broadband** – a type of data transmission that support multiple types of communications sent simultaneously at very high speeds. Broadband is supported by wireless connectivity (cellular network, satellite, microwave, etc.) and also land-based connectivity such as optical fiber.
- **Broadband cellular network** – when you combine the 2, you get a broadband cellular network with the main focuses being (1) connected *cells* or land areas, (2) technologies enabling wireless connectivity, (3) the movement of multiple (perhaps hundreds of thousands) communications simultaneously, and (4) very high speeds.

In general, each mobile phone service provider has its own cellular network. These would include service providers like AT&T, Verizon, and T-Mobile.

Okay, back to the Gs. In short, each new generation in mobile communications brings faster speeds. [1]
- **2G** – 64 kbps (the 1990s, probably before you were born)
- **3G** – 8 mbps (early 2000s, during your childhood)
- **4G** – 50 mbps (came out in 2009)
- **5G** – 10 gbps (introduced in 2018)

Hopefully, you have 5G capability. (At the time I wrote this book, 5G was all the advertising rage, but still not pervasive in the U.S.) The hope (hype?) with 5G is that we can use our wireless devices to enjoy things like high-quality on-demand video streaming and virtual reality streaming with minimal or no latency. **Latency** is the delay that occurs between your action, for example moving your head in a VR game, and the game's response by changing your field of view.

6G

6G is the sixth-generation standard for broadband cellular networks, what every cellular phone company plans to use. 6G will have significantly faster download speeds and really promises to enable the use of technologies like virtual reality, IoT, and AI over a broadband cellular network without any latency or noticeable degradation in speed.

6G is coming, but a couple of years out.

7G... ha! But stay tuned.

Microwave and Satellites

These are communications infrastructure technologies that enable the transmission of large amounts of information and communications simultaneously at extremely high speeds, usually over fairly significant distances. Let's see if we can have a quick and non-weed whacker discussion about these.

Microwaves and **satellites** are line-of-sight communications technologies that transmit massive amounts of data over great distances in a very short period of time. These really are the enabling infrastructure technologies that Internet service providers, TV stations, etc. use to move massive amounts of data around the world.

These are **line-of-sight technologies** because the sender and receiver units have to be able to see each other. That's different from your home WiFi network. Signals inside your home from your wireless router to and from your computer bounce off of walls. So, you can be just anywhere in your house with your computer and still have access to your wireless networks.

Microwaves and satellites use a different transmission methodology, such that the signals don't bounce around. You've seen microwave relay stations and towers. They are positioned about every 8 to 10 miles to enable communications around the curvature of the earth. Satellites are similar except that they are orbiting above the earth. The sending and receiving satellite dishes on the ground must be in line of sight to the satellite.

A great development to watch in this space is Starlink, started by Elon Musk. The goal is to provide Internet service via satellites to the entire global population. There are currently over 1,700 LEOS (low-earth orbit satellites) in the Starlink system, with each connected to dedicated ground transceivers.

Most Innovations Require Multiple Communications Technologies

Suppose, for example, that the "smelling" sensor in your refrigerator detects that the milk is going bad. It could use Bluetooth to communicate that information to your refrigerator's central command center (oh, fancy) which would then use WiFi to connect to your smart home system which would order milk. That order would travel from your WiFi through your 5G network provider to the grocery store. Within your 5G network, communications technologies such as microwaves may be used.

This is an example of the concept of **tethering**. In this case, you're tethering together all these communication technologies to essentially enable your smelling sensor in your refrigerator to communicate with the grocery store.

Wired Technologies

The communications technologies I've covered here are obviously all wireless. And wireless is key to mobility. Remember our discussion of pudding on demand. It was about the new *when*. Mobility, enabled by wireless communications technologies, is all about the new *where*, specifically anywhere.

Of course, wired technologies are still important. Technologies such as fiber optics will continue to play an integral role in the success of technology.

CHAPTER 11

Infrastructure Technologies

Quantum, Cloud, and Nano

So many great and wonderful technologies will be a part of the 4th industrial revolution. In this final chapter, we'll discuss 3 more... quantum computing, cloud technology, and nanotechnology. They don't fit nicely into any of the other specific chapters, mainly because they can be used with any and all the other 4th industrial revolution technologies.

Quantum computing, cloud technology, and nanotechnology are big game changers, enabling unbelievable disruption and innovation.

Quantum computing

Google announced it in October 2019; it had achieved **quantum supremacy**. Its quantum computer solved a mathematical computation in 3 minutes and 20 seconds that would have taken the world's current fastest supercomputer over 10,000 years to solve. [1]

That's the hope and hype around quantum computing... unbelievable speed.

About the time of Google's announcement, the fastest supercomputer in the world, named Fugaku (a joint development of RIKEN and Fujitsu), was capable of performing 415 petaflops. A petaflop is one thousand million million floating point operations per second. **Peta** is the term applied to 1,000,000,000,000,000 or 10^{15}. We measure computer processing speed in terms of how many floating point operations per second (abbreviated as **flops**) the computer can execute. So, Fugaku could perform 415,000,000,000,000,000 operations per second, and – at that rate – it would have taken Fugaku over 10,000 years to solve the mathematical computation that Google's quantum computer did in 3 minutes and 20 seconds.

Like I said, unbelievable speed.

To understand how and why quantum computers are so fast, we need to have a discussion about our classical (current) computer technologies and how they work.

Bits – Classical Computers

All information in classical computing technologies is stored, processed, and communicated in binary form. Binary means two. The smallest unit in binary is called a **bit**, which stands for binary digit. Because binary means two, a bit can take on only one of two values, either 0 or 1. We use this binary concept in many ways... a door/gate is either open or closed, a light is either on or off, a coin flip results in either a head or tail... I think you get the idea.

Inside a computer, there is no single character for the letter P, the number 7, or an asterisk. We represent each of these (and all the other letters, numbers, and special symbols) as a unique series of 0s and 1s.

Time for another question: We (humans) work with numbers in Base 10. So, for each digit position in a number such as 357, there are 10 possible digit representations, 0 through 9. Why do we work in Base 10 and not some other Base scheme such as Base 8 or Base 13? (Sorry no multiple choice here. Think about this for a moment and write your answer below.)

Answer: _____

It's not a trick question, but the simplicity of the answer seems almost too obvious. We work in Base 10 because we have 10 fingers. That's it. If we were 12-fingered mammals, we would be working in Base 12.

Let's consider the number 357. You can break that number down into its Base 10 equivalent as follows:

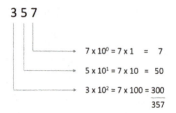

Each digit position can take on 1 of 10 possible representations, 0 through 9. We multiply the digit representation in each position by continually increasing the power of the base, starting with 0 on the right. (I'm sure you remember all of that from your elementary school math classes. Chuckle.)

Binary works the same way, except that each digit position can take on 1 of 2 possible representations, 0 and 1, and the Base is 2. The number 357 in binary would look like this:

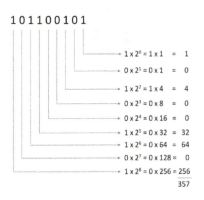

So, when you're working in Excel with the number 357, your computer is working with 101100101. It's quite fascinating when you think about it… the simplicity with which computers represent and work with information.

Qubits – Quantum Computers

But quantum computing is different from classical computing technologies. Quantum computing's most basic unit for representing information is a **qubit**, which stands for quantum binary digit. What's really interesting and distinguishing about a qubit is that it can be both 0 and 1 at the same time. This simultaneous representation of 2 possible states is called *superposition* in quantum mechanics.

So, formally, let's say that a **quantum computer** (or **quantum computing**) is an alternative to classical computing technology that offers unbelievable speed because it performs computations on information in quantum states such as superposition. (There are other quantum states such as interference and entanglement that also offer speed increases and other benefits over classical computing technologies, but it's a bit late to be getting out the old weed whacker.)

The easiest way to think about a qubit is to turn a coin on its side and spin it. While the coin is spinning you can see both heads and tails, the possible states that result from a coin flip. For our discussion, I'll represent a qubit like this:

It looks kind of weird, but the notion is that a qubit can be both 0 and 1 at the same time.

The ability of a qubit to take on 2 values simultaneously is why quantum computing is so much faster than classical technologies. I'll illustrate with an example.

Suppose we wanted to add all possible combinations of 2 classical bits and the same for 2 qubits. It would look like this.

CLASSICAL BITS				QUANTUM BITS			
0 $+1$ 1	0 $+0$ 0	1 $+1$ 10	1 $+0$ 1	1	0 $+0$ 0	10	1
4 Computations				1 Computation			

Using classical bits and technology, you would have to perform addition 4 times. With qubits and quantum computing, you only perform the addition 1 time.

Now, I know you're smart, so I know what you're thinking… "Cool. Quantum computers can do things 4 times faster. Clearly though, 3 minutes and 20 seconds compared to 10,000 years is much more than 4 times faster. I don't get it." (Indeed, you are smart.)

In the example above of adding all possible combinations of 2 classical bits or qubits, the speed increase isn't really 4x, but rather 2^n where n is the number of qubits used. So, if you use 2 qubits, you get an increase of 2^2 in speed (4x). If you then want to add all possible combinations of 3 bits, you can get an increase of 2^3 in speed (8x). Rinse and repeat and you can start to see that speed doubles every time you add another qubit.

This is the notion of **scaling**. We're not adding to or even multiplying speed increases; we are raising the power each time we add another qubit.

Let me illustrate the concept of scaling with a story about rice and the game of chess.

There was once a game maker who invented the game of chess. He took it to the queen, knowing that she loved games. After explaining the game of chess and playing it with the queen for a while, she wanted it and offered to buy it from the game maker.

His price was simple. For the lower left square on the board, pay me 1 grain of rice. For the square next to it on the first row, pay me 2 grains of rice. For the square next to it on the first row, pay me 4 grains of rice. Rinse and repeat until you have paid me for each square, continually doubling the amount of rice payment for each square.

The queen did a quick computation in her head, for the first row:
- Square 1, 1 grain of rice
- Square 2, 2 grains of rice
- Square 3, 4 grains of rice
- Square 4, 8 grains of rice
- Square 5, 16 grains of rice
- Square 6, 32 grains of rice
- Square 7, 64 grains of rice
- Square 8, 132 grains of rice

That was for the first row. To pay for the first row of squares, the queen had to give the game maker only 255 grains of rice, much less than a handful. It sounded like a great deal to her, so she readily accepted the terms of payment.

Bad decision. Why? Because of scaling.

Think about the second row of 8 squares. The rice payment for each square is 256, 512, 1024, 2048, 4096, 8192, 16,384, and 32,768. We're already at 32,768 grains of rice for a single square, and we've got 6 more rows of 8 squares each to go.

I don't think we need to go through all the remaining squares. But I will tell you this, in the end the queen had to pay for the chess game an amount of rice equivalent in size to Mount Everest.

That's the notion of scaling.

So, with each new qubit, we double the speed of quantum computing.

The Medical Realm

Medicine, health care, medical… whatever you want to call it… is perhaps the field of study and application that will benefit most from advances in 4th industrial revolution technologies. And quantum computing will dramatically impact and disrupt (in a good way) the field of medicine.

For example, the human genome – all genetic information in a person – contains 3 billion base pairs. Storing this information using classical bits requires approximately 1.5 gigabytes of storage. That same amount of information can be stored in 34 qubits. And, if we double the number of qubits to 68, we would have enough room to store the complete genome of *every* person in the world. [2]

Working with and researching the human genome are among the most complicated and time-intensive tasks not only in medicine but in everything we use computers for. Drug testing and trials are another area that will benefit from advances in quantum computing. Prior to receiving approval for human testing, pharmaceutical companies must run millions and millions of "simulations" to understand drug interactions, side effects, and so on. Sometimes this can take years. But, with quantum computing…

The Artificial Intelligence Realm

As you learned in the chapter on AI, many AI (1) use very large and complex sets of information to arrive at an answer, and/or (2) rely heavily on learning from vast amounts of information.

Using quantum computing, we can reduce the size of the information for arriving at an answer because we can store and process that information in qubits. The less information you have to consider for making a decision, the faster you can make the decision (generally).

And, we can greatly reduce the time it takes to train an AI, i.e., its learning. Instead of using classical bits to feed the AI one scenario at a time, we can simultaneously feed the AI thousands, if not millions, of scenarios at once using qubits.

The Security Realm – A Rather Serious Implication

While quantum computing promises great computational speed, there exist some significant downsides, especially in the security realm.

In the cryptocurrency and blockchain chapter, we talked briefly about the 256-bit hash or key value that uniquely describes each block. This is an important security feature of blockchain technologies that is also used for the private keys you receive that identify you are the owner of cryptocurrency.

A 256-bit key has 115,792,089,237,316,195,423,570,985,008,687,907,853,269, 984,665,640,564,039,457,584,007,913,129,639,936 possible combinations. It's a really, really big number. Even the fastest of today's classical computers would take millions of years to try all possible combinations to "hack" the code.

But, if we build a quantum computer that can work with 256 qubits simultaneously, a hacker could conceivably try all possible combinations at once. Yikes!

Are We Going to Get There?

The answer is most certainly yes. We are just now beginning to see great advances in the space of quantum computing, so it's going to take a while, but we will get there.

For example, Chinese startup company SpinQ offers a desktop quantum computer for $5,000. It can work with only 2 qubits. "Desktop" is a relative term her, as the quantum computer weighs a hefty 121 pounds. [3]

As of early 2022 when I wrote this book, Google was working on Bristlecone, a quantum computer capable of working with 72 qubits.

Some of the big commercial competitors in this space include Honeywell, IBM, Google, Amazon, Microsoft, Rigetti, and Fujitsu. As a side note, IBM is indeed a big competitor in this space. Shortly after Google announced its quantum supremacy achievement in 2019, IBM refuted it, saying that the magnitude was nowhere near the scale that Google had announced. [4]

Government initiatives/players in this space include:
- National University of Defense Technology, China (government security applications)
- National Supercomputer Center in Wuxi, China (climate science)
- Lawrence Livermore National Laboratory, U.S. (US National Nuclear Security Administration modeling and simulations)
- U.S. Department of Energy, U.S. (energy efficiency, genetic applications)
- RIKEN Center for Computational Science (location of Fugaku), Japan (drug discovery, personalized medicine, climate forecasting, clean energy development)

Amazon has announced quantum-computing-as-a-service through Amazon Web Services (AWS). [5]

▲ ▲ ▲

Cloud Computing (As-a-Service)

(Cloud computing has been around for upwards of 30 years. So, it's definitely not a 4th industrial revolution technology. But, the notion of "cloud" storage and processing are central to the success of innovations in the 4th industrial revolution.)

In the simplest of terms, **cloud computing** is the storing and accessing of information and software on the Internet instead of your computer's hard drive. You're probably already "doing" cloud computing. If you use Dropbox for storing and sharing files, you're doing cloud computing. iCloud and Google Drive are examples of cloud computing.

Even things like Spotify are examples of cloud computing. With Spotify you can stream music from the Internet without having to store the songs on a device. (You can, of course, download your playlists from Spotify.) Netflix is also an example of cloud computing. While you can download movies from Netflix for offline viewing, you don't have to.

Cloud computing has become synonymous with "as-a-service." As-a-service is the opposite of ownership. Again, think about Spotify. You can access all the music you want as-a-service without paying for the actual ownership of the music. As long as you pay your monthly subscription, you can listen to all the music you want. Netflix is the same. Pay a monthly fee and watch all the moves you want. Of course, when you drop Netflix, you no longer have access to the movies because you didn't pay to own them.

The same is true for cloud computing. Instead of paying for ownership, you pay a monthly fee to have access.

Let me give you a simple example. Let's say you want to buy a thumb drive to back up your computer. You have no intentions of ever accessing the data on the thumb drive unless your computer crashes or gets stolen. So, you've decided to buy a 2TB (2 terabyte) thumb drive. That will cost you about $40. Small price to pay for back up, right? Indeed, but it can be even less in price if you use Amazon Web Services (AWS) and its S3 Glacier Deep Archive storage-as-a-service in the cloud. Using that service, 2TBs of back up data will cost you $.99 per month.

So, the equivalency is simple to determine, $40 now for a thumb drive or $.99 per month for 40 months (3 years and 4 months) spread out over that time period.

Of course, you can argue that thumb drives are going to get cheaper in price over time (and that would be true), but that would imply that you would buy another thumb drive whenever they dropped in price. Now, you've spent $40 on the original thumb drive and perhaps another $30 18 months down the road when the price drops. Doesn't make sense.

As-a-service shows up in all aspects of our lives. People living in a big city like New York City often opt to not own a car but rather use transportation-as-a-service. If you own a car in New York City, you have all the typical costs of ownership… maintenance, parking, license plates, gas, insurance, etc. Or, you can opt to pay for transportation – taxi cab, subway, Uber, etc. – when you need it and <u>only</u> when you need it.

Break-Even Analysis: The Business Decision of Cloud Computing

The business decision of cloud computing is simple and no different from your personal decision of transportation-as-a-service. Do you want to handle all the maintenance and other costs associated with the car, or do you want someone else to do so? Do you want to constantly pay for the "latest and greatest," or do you want to have access to today's best? And, simply put, is it cheaper to buy and own or to "rent" what you need when you need it?

As you learn more about cloud computing from a business point of view, you'll see these 3 terms:
- *Infrastructure-as-a-service (IaaS)* – basic technology needs like virtual servers and storage.
- *Platform-as-a-service (PaaS)* – a step up from IaaS that also includes development environments and tools. For example, you may have applications you build for Windows operating systems (Microsoft), but you want to build also in the UNIX space.
- *Software-as-a-service (SaaS)* – the use of application software you need but don't want to pay for the ownership and associated costs.

Of the three, SaaS is the dominant model, accounting for over half of all cloud computing spending by businesses. [6] A good example is Salesforce. It was the first big player in the SaaS model for customer relationship management (CRM) software.

Cloud Providers

There are many providers in the cloud computing space. If you're interested in looking at any of them, consider the list below. [7]

- Amazon Web Services (AWS)
- Microsoft Azure
- Google Cloud
- Alibaba Cloud
- IBM Cloud
- Oracle
- Salesforce
- SAP
- Rackspace Cloud
- VMWare

Again, there are many.

▲ ▲ ▲

Nanotechnology

Okay, this is going to be a very short and sweet, non-weedy discussion. It's really challenging to provide a concise, yet approachable, definition of nanotechnology, so let's just chat.

Nanotechnology is all about extremely small things, and using and manipulating those extremely small things to manufacture larger things. Incidentally, these "larger" things are also extremely small. When I think of nanotechnology, I think of it as 3D printing on steroids, lots of steroids. Recall that subtractive manufacturing is taking some sort of material and whittling, slicing, and compressing it to make something smaller. I talked about trees before and making 2x4s, matches, and toothpicks from them. That's subtractive manufacturing.

Additive manufacturing is about building from the bottom up. In 3D printing, you manufacture something by continually adding layer upon layer of material until your product is complete.

The same is true for nanotechnology, but the "layer" becomes exceedingly small.

Nano stands for one-billionth. Not one billion, but rather one-billionth. So, 1 nanometer is one-billionth of a meter. There are just over 25 million nanometers in an inch.
- A human hair is about 100,000 nanometers in width.
- A sheet of paper is about 100,000 nanometers thick.
- Your fingernails grow about 1 nanometer per second. (How often do you have to cut your fingernails?)
- Your hair also grows about 1 nanometer per second. So, your hair grows about 1 inch in 25 million seconds, or roughly 289 days. (That's, of course, the average. Yours may grow faster or slower.)
- The popular comparative scale: If a marble were a nanometer, then one meter would be the size of the earth. (Wrap your brain around that one.)

Hopefully, that gives you some sort of perspective for the scale of nanotechnology.

Formally, **nanotechnology** is the manipulation of material with at least one dimension of size between 1 and 100 nanometers. That's really, really, really small. Something that is 100 nanometers in width is 1,000 times smaller than a human hair. Like I said, really, really, really small. It takes unbelievably complex and expensive equipment to manufacture things on that scale of size.

But the possibilities of nanotechnology can go even smaller. Think about being able to build things on a molecule by molecule basis, or atom by atom basis. In case you've forgotten, atoms are single neutral particles. Molecules are neutral particles made up of two of more atoms bonded together. (That 7th grade general science class sure is coming in handy now, isn't it?)

It's theoretically possible using nanotechnology to build a computer – RAM, CPU, battery power supply, etc. – that would be about the width of a human hair.

Perhaps the most far-reaching and beneficial innovations of nanotechnology will be in the field of medicine. For example, researchers are developing nanoparticles to deliver drugs, heat, light, and other substances to specific types of cells. The nanoparticles are engineered in such a way that they are only attracted to certain types of diseased cells, such as cancer cells. In doing so, the nanoparticles would direct treatment only to the cancer cells, without damaging any of the surrounding healthy cells. [8]

Other researchers are working on building nanorobots that could be programmed to repair diseased cells. These nanorobots would emulate the antibodies of our natural healing process. [9]

Companies and Initiatives to Watch

- Applied Materials
- Avano
- CENmat
- CMC Materials
- DuPont de Nemours
- NanoScientifica
- Peak Nano

- Taiwan Semiconductor
- Tandem Nano
- Thermo Fisher Scientific

I'm going to stop there. If you're really interested in nanotechnology, I would encourage you to explore the on-going research into the application of nanotechnology in medicine. Truly, truly fascinating.

▲ ▲ ▲

THE TAKEDOWN

Meanderings Of An Old Man

If I Were in Your Shoes

(3D-Printed Shoes, That Is)

Alright, lets' wrap up our time together by continuing to look toward the future.

Is Technology in Your Crystal Ball?

We've covered some important sets of topics, including:

- The Internet of (All) Things
- Cryptocurrency & Blockchain
- Artificial Intelligence
- Extended Reality
- 3D Printing
- Autonomous Vehicles
- Drones
- Energy
- Batteries
- Wireless Communications Technologies
- Sensing Technologies
- Quantum Computing
- Nanotechnology
- Cloud Computing

It's been easy for me to make the case for each as being important in the 4th industrial revolution. From a career perspective, they are equally important as well.

I did a quick search of important skill sets for the future and found some interesting results.

Top Tech Skills for the Future of Work (medium.com [1])

Top skills in high demand now and at least for the next 5 to 10 years include:
1. Artificial intelligence and machine learning
2. Data science and analytics
3. Blockchain
4. Cloud computing/DevOps
5. Digital marketing
6. Full stack development
7. Cyber security

The Most In-Demand Tech Skills (learntocodewith.me [2])

The top 18 digital skills to learn in 2022 include:
1. Artificial intelligence
2. Machine learning
3. Data science and analytics

4. Data engineering
5. Data visualization
6. Cybersecurity
7. Cloud computing/AWS
8. Extended Reality
9. Internet of Things
10. UI/UX design
11. Mobile development
12. Blockchain
13. Quantum computing
14. Robotics
15. Product management
16. Salesforce/CRMs
17. Low-code platforms
18. Programming languages in general

Guide to Future-Oriented Skills (hrforecast.com [3])

The highest paying skills in technology include:
1. Cloud computing
2. Cybersecurity
3. Artificial intelligence and machine learning
4. Big data analytics
5. Extended reality
6. Blockchain
7. Video production
8. User experience

In-Demand Tech Skills for Technology Careers (indeed.com [4])

The top tech skills that are in demand include:
1. Artificial intelligence
2. Cybersecurity
3. Cloud computing
4. Software engineering
5. Software development
6. Project management
7. UI/UX design
8. Data analytics
9. Data science

10. Programming
11. Computer graphics
12. Digital marketing
13. Translation
14. Mobile development
15. Virtualization
16. Network maintenance
17. Business analysis
18. People management
19. Technical writing
20. Blockchain
21. Quantum computing
22. Robotics
23. Video production
24. Game development
25. Industrial design

Top Tech Skills Growing the Fastest in 2021 (forbes.com [5])

The fastest growing (and most lucrative) tech skills include:
1. Quantum computing
2. Connected technologies (IoT)
3. Fintech
4. Artificial intelligence and machine learning
5. IT automation
6. Natural language processing
7. Cybersecurity
8. Software development methodologies
9. Cloud technologies
10. Parallel computing

In-Demand IT Skills for 2021 (itcareerfinder.com [6])

And finally, the 10 top in-demand IT skills include:
1. Cybersecurity
2. Big data and Internet of things
3. Artificial intelligence and machine learning
4. Cloud computing
5. Software development
6. Robotics process automation

7. Project management
8. Autonomous driving
9. IT services management
10. Marketing automation

Are you beginning to see some trends? The future looks very bright for tech specialists in a variety of the technology sets we've discussed including, artificial intelligence (and machine learning), the Internet of (all) things, blockchain, cloud computing, quantum computing, extended reality, and autonomous vehicles.

If you want to get into tech, things are looking very, very good.

But, even if you don't want to get into tech, the lists above are important because they help you to see the big picture of where the world is going. No doubt, things like artificial intelligence will be integrated into every aspect of your life in the future. The same is true for blockchain and many of the other 4th industrial revolution technologies.

Even if you don't want to build the tech, the tech will still very much be a part of your life.

▲ ▲ ▲

Moving Forward

Answer this Question: In the Future, Will You Control the Technology, or Will the Technology Control You?

I don't really know how much explanation you need for that question. I'm betting you understand it well.

If you know something about how the technology works and why organizations are using it, you're more likely to be in control of the technology, as opposed to the other way around.

Find Schools with a Futuristic and 4th Industrial Revolution Focus

I'm a big believer in learning and education. And those don't necessarily have to come from a typical 4-year college or university. You can go the careers in technical education (CTE) route or even strike out on your own to build your business and dream.

Regardless, find learning resources that focus on the future and the 4th industrial revolution technologies.

If you do go the school route, try to find schools that are focusing some of their teaching on the future. Keep in mind that you're not going to school to get a job, you're going to school to prepare for a life-long career.

Search for the Holy Donut Hole

The donut (with the hole) was invented in the mid 1800s. Oddly enough, it wasn't until the mid 1900s that someone finally decided to start selling donut holes. Almost 100 years passed of making donuts before someone got the idea to fry, glaze, and sell the hole.

Donut holes are everywhere, waiting for you to find them.

Invest on the Edge

Start watching up-and-coming companies in the 4th industrial revolution space. Invest some early money in them.

If you had purchased $500 of Amazon stock during its IPO (initial public offering) in 1997, you would have almost $1 million today (using early 2022 stock price of $2,800 per share).

Not a bad return. 2,000x return in 25 years.

Grain of salt… many people invested $500 way back then in other (now defunct) companies and lost it all. Invest small amounts, across several companies.

Make Passive Income

There are 2 types of income, active and passive. Active income is what you receive for doing a job. You go to work, you get paid. You don't go to work, you don't get paid. Active income is what most people do… get a job and get paid to do that job.

Passive income is about making money *while you sleep*. Investing in the stock market, playing the wild-and-crazy crypto market, and owning rental real estate are all examples of passive income. Be careful, passive income can be at a loss. Playing the crypto market can make you a lot of money; on the flipside, you can lose every bit of your investment.

As you approach your future, find ways to build an income portfolio of both active and passive income.

Creact – Be Creative and Take Action

Creact is a weird combination of being **CRE**ative and taking **ACT**ion. Creact really means, instead of just taking the "traditional" action that everyone else takes, be creative and take new and different action.

That's entirely possible using 4th industrial revolution technologies. Using those technologies, we get to do things in an entirely different way than before.

Out with the old, in with the new. It's a mindset.

Learn One Simple – Yet Powerful – Math Concept

Whether you're thinking about your retirement nest egg, trying to rectify the world's environmental problems, or even trying to lose weight, there is one simple math concept that will help you.

Let me tell you a story. There was a woman, very successful, who made millions of dollars in her lifetime as an entrepreneur. A brave young 12-year-old approached her and asked, "Will you tell me how to become a millionaire?" She loved the young man for his courage and replied, "Of course. Tell me, do you know how to make a dollar, just one dollar?" The boy quickly answered, "Sure. I have a lot of small businesses like raking leaves and walking dogs. I can make a dollar right now."

To which the woman replied, "Good, go do that a million times."

The math concept…

A small number times a big number is a big number.

Think about Walmart. It generates an annual revenue in the neighborhood of $500 billion. That's almost a half-trillion dollars. (Big number.) If Walmart could use 4th industrial revolution technologies like drones and IoT to save just one more penny (small number) on every dollar of revenue, Walmart would save $5 billion per year. A small number times a big number is a big number.

There are 6.6 billion smartphones in the world today. Depending on the make and model, each costs about $1.75 per year to charge, if you completely drain the battery of your smartphone daily and charge it fully every night.

Now, $1.75 doesn't sound like a lot of money. But, think in terms of a small number times a big number is a big number. Given that there are 6.6 billion smartphones, the total annual cost to recharge all those phones is somewhere in the neighborhood of $10 billion.

Now that's a big number. What if we could all use something like solar energy to recharge our smartphones? We would collectively save $10 billion annually, and we would also be using solar energy instead of fossil fuel-based energy. Seems like a good way to save the environment.

A small number times a big number is a big number.

▲ ▲ ▲

And on a Personal Note

Get and Give (aka, Buy One, Give One or BoGo)

One of the (many) things I admire about your generation is the notion of helping others while helping yourself. As I tell my students in the classroom, "You have two hands. One for helping yourself, and one for helping others." Your generation is particularly keen on doing this, and I applaud you for that.

One of the simplest ways to achieve this is to participate in BoGo programs, not buy-one-get-one but rather buy-one-give-one. Some of those programs include:

- One World Play Project – soccer balls. (https://www.oneworldplayproject.com/)
- Bixbee – school backpack with supplies. (https://bixbee.com/)
- Bombas – t-shirt, socks, or underwear. (www.bombas.com)
- Roma – boots. (www.romaboots.com)
- Smile Squared – toothbrushes. (https://www.ishopwithpurpose.com/smile-squared)
- Soap Box – bars of soap. (https://www.soapboxsoaps.com/)
- Figs – medical scrubs. (www.wearfigs.com)
- Better World Books – books. (www.betterworldbooks.com)
- Warby Parker – eye wear. (www.warbyparker.com)
- BogoLight – flashlights and solar-powered lights. (www.bogolight.com)

That's just a few of the many. Find your passion and participate.

Just Give

Lots of wonderful opportunities here.

My personal favorite is PocketChange (https://pocketchange.social/). PocketChange is social media with a purpose. Get the app. Every like, post, and reply you make results in free money sent to a charity at no cost to you.

I love one of its statements in About Us… "The more time you spend on PocketChange, the better the world gets – it's social media without the guilt trip."

Find Bigfoot

> (I could have chosen finding the pot of gold at the end of the rainbow, the fountain of youth, or anything else for that matter. I chose Bigfoot because he/she is a personal favorite of mine.)

The world is full of fanciful unknowns and wild stories… Bigfoot, Nessie, aliens, Yeti. We often refer to these as urban legends.

Go on a journey to find one. The end goal isn't the "find," but rather the journey.

Chase a Dream, Every Dream

Every dream you have deserves some of your time. Want to write a book, give it a try. Want to run a marathon, give it a try.

Like finding Bigfoot, the journey is what matters.

Don't Inherit the World from Your Parents; Prepare It for Your Children

Think forward, not backward.

Yes, you inherited this world (and all its problems) from my generation. I just saw that the U.S. national debt exceeded $30 trillion for the first time ever. We poked holes in the ozone layer with huge cans of aerosol spray. We used unbelievable amounts of fossil fuels without regard for the environment.

It's the hand of cards you've been dealt.

Your focus should be on making the world a better place for the generations to come after you.

Try to Put Your Socks on Backward (Heel First)

I love this one, because it makes no sense at all. You can't put your socks on heel first.

But DoorDash and Uber Eats made no sense at first, either. Online grocery shopping defied conventional wisdom. Currency without government backing. Water bottles that can diagnose the flu. They were all like trying to put your socks on backward, at first.

Now… perhaps maybe we should rethink socks.

▲ ▲ ▲

My Final Thought... Be Yourself

People are happiest when they are themselves, when they don't have to hide behind the "closet door," when they can be completely, open, and honest with the world around them.

I realize that probably right now you're thinking that you've got a parent or two constraining (for lack of a better term) this. And that's okay. Parenting is the most difficult job in the world, finding that delicate balance between letting you fly and watching over you in the nest.

But ultimately, you have to be happy with yourself and who you are – physically, mentally, emotionally, and spiritually.

I hope to meet the <u>real</u> you someday.

▲ ▲ ▲

Acknowledgments

In 2019, we began the work of crafting *The Fourth Industrial Revolution* course at the University of Denver. It is now required of all business majors and minors. It was my privilege and pleasure to work with a number of great faculty in bringing that course to life. They include Joshua Ross, Aaron Duncan, Tamara Hannaway, Tom Rankin, Gisella Bassani, Aaron Duncan, and Colleen Reilly.

I'm also privileged to have great friends who lend their expertise to and support for my writing efforts. They include Steven Hartley, Michael Myers, and Curt Coffman.

Finally, I could never complete such an undertaking without the unwavering support of my family. Katrina and Alexis make me smile every day. Darian and Trevor helped review the manuscript. My wife Pam is my rock. I love you all.

Trademarks and Name Usage

Throughout this book, I've named over 1,000 people, products, and organizations. My use of them does not in any way imply an endorsement on their part for the contents of this book. As applicable, the trademarks and all other intellectual property are the properties of the organizations.

Art Work – VectorStock

The art work in this book has been provided by Latte (cover and interior design) and many wonderful artists at VectorStock. Latte developed the pieces of art unique to the book. The artists of VectorStock provided the rest of the line art.

VectorStock is a great web resource for obtaining line art. A wealth of great artists, tons of images to choose from, and very, very reasonable prices. VectorStock provided the following images.

Pudding, page 1 Designed by ZdenekSasek (Image #36561643 at VectorStock.com)
Farmland, page 11 Designed by crop_ (Image #15069351 at VectorStock.com)
Old telephone, page 13 Designed by suricoma (Image #15149239 at VectorStock)
Weed Whacker, page 22 Designed by Ylivdesign (Image #39441858 at VectorStock.com)
The Internet of Things, page 25 Designed by defmorph (Image #23867242 at VectorStock.com)
Future, appearing in many places Designed by tentacula (Image #22673825 at VectorStock.com)
Bitcoin, page 43 Designed by longquattro (Image #20168951 at VectorStock.com)
Ethereum, page 43 Designed by longquattro (Image #20168951 at VectorStock.com)

Artificial Intelligence, page 69 Designed by defmorph (Image #19788510 at VectorStock.com)

Extended Reality, page 105 Designed by tentacula (Image #8605341 at VectorStock.com)

3D Printing, page 125 Designed by antto (Image #14531120 at VectorStock.com)

Motorcycle, page 127, Designed by Candtp123 (Image #1110118 at VectorStock.com)

Jacket, page 128, Designed by ollymolly (Image #13221761 at VectorStock.com)

House, page 130, Designed by ilyaka (Image #1075440 at VectorStock.com)

Cupcake, page 131, Designed by apokusay (Image #13673585 at VectorStock.com)

Prosthetic Limb, page 133, Designed by MiaAkimo (Image #26038372 at VectorStock.com)

Bicycle Helmet, page 134, Designed by nickylarson974 (Image #21964521 at VectorStock.com)

3D Printed, page 137, Designed by bizart (Image #10447494 at VectorStock.com)

Toothbrush, page 141, Designed by bioraven (Image #2092577 at VectorStock.com)

Baby Shoes, page 129 Designed by ToscaDigital (Image #35514389 at VectorStock.com)

Autonomous Vehicle, page 147 Designed by tentacula (Image #38293578 at VectorStock.com)

Car Rear-ending, page 155 Designed by leremygan (Image #26128996 at VectorStock.com)

Truck, page 157 Designed by CMYK (Image #9474884 at VectorStock.com)

Drone, page 165 Designed by talashow (Image #17829099 at VectorStock.com)

Energy, page 183 Designed by tentacula (Image #9547557 at VectorStock.com)

Solar, page 190 Designed by vikivector (Image #23486645 at VectorStock.com)

Wind Farm, page 192 Designed by angelha (Image #22970261 at VectorStock.com)

Batteries, page 196 Designed by infadel (Image #32441147 at VectorStock.com)

Senses, page 201 Designed by fourleaflover (Image #38426974 at VectorStock.com)

Communications Technology, page 216 Designed by Vectorchoice (Image #9363252 at VectorStock.com)

RFID, page 217 Designed by Tartila (Image #34943477 at VectorStock.com)

Whether you're writing a book, preparing an industry report, or just a presentation for class, I highly recommend VectorStock.

Notes

The Setup

1. Hoover, Gary. "Largest Companies 1917 and 1929." *The American Business History Center*, January 15 2022, https://americanbusinesshistory.org/largest-companies-1917-and-1929/.
2. "Largest Companies by Market Cap." *CompaniesMarketCap*, https://companiesmarketcap.com/.

Chapter 1

1. Lee, Ahyoung & Wang, Xuan & Nguyen, H. & Ra, Ilkyeun. (2018). A Hybrid Software Defined Networking Architecture for Next-Generation IoTs. KSII Transactions on Internet and Information Systems. 12. 932-945. 10.3837/tiis.2018.02.024.
2. Watters, Ashley. "30 Internet of Things Stats & Facts for 2022." *CompTIA*, February 10 2022, https://connect.comptia.org/blog/internet-of-things-stats-facts.
3. Trafton, Anne. "Sensors Woven into a Shirt Can Monitor Vital Signs." *MIT News | Massachusetts Institute of Technology*, MIT News Office, https://news.mit.edu/2020/sensors-monitor-vital-signs-0423.
4. Fisher, Tim. "What Is a Smart Bed?" *Lifewire*, February 3 2022, https://www.lifewire.com/smart-bed-4161313.
5. Sakharkar, Ashwini. "Printing Sensors Directly on Human Skin." *Tech Explorist*, October 9 2020, https://www.techexplorist.com/printing-sensors-directly-human-skin/35675/
6. Ablondi, Bill, and Jack Narcotta. "Post-COVID Smart Home Device Markets Set to Rebound in 2021." *Strategy Analytics*, July 14 2020, https://news.strategyanalytics.com/press-releases/press-release-details/2020/Strategy-Analytics-Post-COVID-Smart-Home-Device-Markets-Set-to-Rebound-in-2021/default.aspx
7. Krebs, Brian. "Target Hackers Broke in via HVAC Company." *Krebs on Security*, February 5 2014, https://krebsonsecurity.com/2014/02/target-hackers-broke-in-via-hvac-company/
8. "Global IoT spending to break $1 trillion by 2023 – fueled by solid consumer and commercial adoption." *IoTnews*, June 14 2019, https://iottechnews.com/news/2019/jun/14/global-iot-spending-break-1-trillion-2023-fuelled-solid-consumer-and-commercial-adoption/
9. Lueth, Knud Lasse. "State of the IOT 2020: 12 Billion IOT Connections, Surpassing Non-IoT for the First Time." *IoT Analytics*, November 8 2021, https://iot-analytics.com/state-of-the-iot-2020-12-billion-iot-connections-surpassing-non-iot-for-the-first-time/
10. Williams, Robert. "Oral-B Connects AI-Powered Toothbrush to Mobile Apps for Personalized Tips." *Marketing Dive*, February 27 2019, https://www.marketingdive.com/news/oral-b-connects-ai-powered-toothbrush-to-mobile-apps-for-personalized-tips/549182/
11. Truong, Jessica. "How to Hack Self-Driving Cars: Vulnerabilities in Autonomous Vehicles." *HackerNoon*, July 13 2021, https://hackernoon.com/how-to-hack-self-driving-cars-vulnerabilities-in-autonomous-vehicles-jh3r37cz
12. "Industrial Internet of Things Market Size Worldwide from 2017 to 2025." *Satista*, Martin Placek, March 10 2021, https://www-statista-com.du.idm.oclc.org/statistics/611004/global-industrial-internet-of-things-market-size/

Chapter 2

1. Prosser, Emmett. "Aaron Rodgers to take a part of his 2021 salary in Bitcoin, give $1 million of digital currency to fans." *USA Today*, November 1 2021, https://www.usatoday.com/story/sports/nfl/packers/2021/11/01/packers-aaron-rodgers-takes-portion-his-salary-bitcoin/6244838001/.
2. Foxley, William "Why the Navajo are mining Bitcoin." *Bitcoin Magazine*, November 4 2021, https://bitcoinmagazine.com/culture/why-the-navajo-are-mining-bitcoin.

3. Sigalos, MacKenzie. "Incoming Ney York mayor Eric Adams vows to take first three paychecks in bitcoin." *CNBC*, November 4 2021, https://www.cnbc.com/2021/11/04/new-york-mayor-elect-eric-adams-to-take-first-3-paychecks-in-bitcoin.html.
4. NBC 6, "Miami to Distribute 'Bitcoin Yield' to Residents: Mayor Suarez." *NBC 6 South Florida*, November 11 2021, https://www.nbcmiami.com/news/local/miami-to-distribute-bitcoin-yield-to-residents-mayor-suarez/2618085/.
5. Bumbaca, Chris. "Trevor Lawrence partners with Blockfolio, will have signing bonus placed into cryptocurrency account." *USA Today*, April 26 2021, https://www.usatoday.com/story/sports/nfl/draft/2021/04/26/trevor-lawrence-jaguars-signing-bonus-cryptocurrency/7383149002/.
6. "Global Blockchain Market Report: Market Size is Projected to Grow from $4.9 Billion in 2021 to $67.4 Billion by 2026." *GlobalNewswire*, December 16, 2021, https://www.globenewswire.com/news-release/2021/12/16/2353499/28124/en/Global-Blockchain-Market-Report-2021-Market-Size-is-Projected-to-Grow-from-4-9-Billion-in-2021-to-67-4-Billion-by-2026-at-a-CAGR-of-68-4.html.
7. Zagorsky, Jay. "This country has just made bitcoin legal tender. Here's what it means." *World Economic Forum*, September 7 2021, https://www.weforum.org/agenda/2021/09/bitcoin-legal-tender-el-salvador-economics-finance/.
8. "It costs 2¢ to make a penny and 7¢ to make a nickel, but CENTS Act could bring those costs down." *GovTrack Insider*, July 8 2019, https://govtrackinsider.com/it-costs-2-to-make-a-penny-and-7-to-make-a-nickel-but-cents-act-could-bring-those-costs-down-aa6aabfc9a8b.
9. de Best, Raynor. "Companies that accept cryptocurrency in the U.S. as of March 9, 2021, by city." *Statista*, March 25 2021, https://www.statista.com/statistics/1223979/firms-with-crypto-payment-solution-usa-city/.
10. Sephton, Connor. "How many cryptocurrencies are there?" *Currency.com*, January 27 2022, https://currency.com/how-many-cryptocurrencies-are-there#:~:text=attention%20%E2%80%93%20Photo%3A%20Shutterstock-,How%20many%20cryptocurrencies%20are%20there%3F,more%20than%2012%2C170%20in%20existence.
11. www.coinmarketcap.com.
12. Bharathan, Vipin. "US Postal Service Files A Patent For Voting System Combining Mail And A Blockchain." *Forbes*, September 20 2020, https://www.forbes.com/sites/vipinbharathan/2020/09/20/us-postal-service-files-a-patent-for--voting-system-combining-mail-and-a-blockchain/.
13. Henderson, James. "De Beers tracks 100 diamonds through supply chain using blockchain." *SupplyChain*, May 17 2020, https://supplychaindigital.com/technology/de-beers-tracks-100-diamonds-through-supply-chain-using-blockchain, Supple Chain Digital.
14. Vitasek Kate, Bayliss John, Owen Loudon, Srivastava Neeraj. "How Walmart Canada Uses Blockchain to Solve Supply-Chain Challenges." *Harvard Business Review*, January 5 2022, https://hbr.org/2022/01/how-walmart-canada-uses-blockchain-to-solve-supply-chain-challenges.
15. Daley, Sam. "19 Blockchain Companies Boosting the Real Estate Industry." *Builtin*, July 30 2021, https://builtin.com/blockchain/blockchain-real-estate-companies.
16. Daley, Sam. "How Using Blockchain in Healthcare is Reviving the Industry's Capabilities." *Builtin*, July 30 2021, https://builtin.com/blockchain/blockchain-healthcare-applications-companies.
17. Hydrogen. "5 Common Blockchain Applications in Financial Services." *Hydrogen*, December 12 2019, https://www.hydrogenplatform.com/blog/5-common-blockchain-applications-in-financial-services.
18. "The Unbanked." *Findex*, 2017, https://globalfindex.worldbank.org/sites/globalfindex/files/chapters/2017%20Findex%20full%20report_chapter2.pdf.
19. Mearian, Lucas. "10 top distributed apps (dApps) for blockchain." *ComputerWorld*, December 30 2019, https://www.computerworld.com/article/3510457/10-top-distributed-apps-dapps-for-blockchain.html.

20. Locke, Taylor. "What are DAOs? Here's what to know about the 'next big trend' in crypto." *CNBC*, October 25 2021, https://www.cnbc.com/2021/10/25/what-are-daos-what-to-know-about-the-next-big-trend-in-crypto.html.

21. Locke, Taylor. "What are DAOs? Here's what to know about the 'next big trend' in crypto." *CNBC*, October 25 2021, https://www.cnbc.com/2021/10/25/what-are-daos-what-to-know-about-the-next-big-trend-in-crypto.html.

22. Commisso, Danielle. "PayPal Now Sells Cryptocurrency, but Are People Buying?" *CivicScience*, February 9 2021, https://civicscience.com/paypal-now-sells-cryptocurrency-but-are-people-buying/.

23. Browne, Ryan. "More than $90 million in cryptocurrency stolen after a top Japanese exchange is hacked." *CNBC*, August 19 2021, https://www.cnbc.com/2021/08/19/liquid-cryptocurrency-exchange-hack.html.

24. https://ycharts.com/

25. Adams, Michael. "The 7 Best Stablecoins to Buy." *U.S. News*, Jan 4 2022, https://money.usnews.com/investing/cryptocurrency/slideshows/what-is-the-best-stablecoin-list.

26. Hale, Jacob. "Top 10 most expensive NFTs ever sold." *Dexerto*, February 17 2022, https://www.dexerto.com/tech/top-10-most-expensive-nfts-ever-sold-1670505/.

27. Khatri, Amara. "Snoop Dogg To Sell 1000 Passes For Private Metaverse Party." *Bitcoin Insider*, September 24 2021, https://www.bitcoininsider.org/article/127968/snoop-dogg-sell-1000-nft-passes-private-metaverse-party.

28. Robinson, Randy. "Snoop Dogg's Metaverse Neighbors Paid $1.23 million for Digital Real Estate." *Merry Jane*, December 7 2021, https://merryjane.com/news/snoop-doggs-metaverse-neighbors-paid-dollar123-million-for-digital-real-estate.

29. "Best NFT Marketplaces." *The Ascent*, February 17 2022, https://www.fool.com/the-ascent/cryptocurrency/nft-marketplaces/.

Chapter 3

1. Brown, Mike. "A Google Algorithm Was 100 Percent Sure That a Photo of a Cat Was Guacamole." *Inverse*, Inverse, June 20 2019, https://www.inverse.com/article/56914-a-google-algorithm-was-100-percent-sure-that-a-photo-of-a-cat-was-guacamole

2. *Artificial Intelligence Market Size Analysis Report, 2021-2028*. Grand View Research, June 2021, https://www.grandviewresearch.com/industry-analysis/artificial-intelligence-ai-market

3. "Sophia." *Hanson Robotics*, September 1 2020, https://www.hansonrobotics.com/sophia/

4. "Kismet, the Robot." *MIT Education*, http://www.ai.mit.edu/projects/sociable/baby-bits.html

5. Likens, Scott, et al. *AI Predictions 2021*. PwC, 10 February 2017, https://www.pwc.com/us/en/tech-effect/ai-analytics/ai-predictions.html

6. Malaviya, Sandip. "Top 10 AI Development Tools, Frameworks for 2021." *Samarpan Infotech*, Samarpan Infotech, September 29 2021, https://www.samarpaninfotech.com/blog/best-ai-development-tools-frameworks/

7. Daley, Sam. "28 Examples of Artificial Intelligence Shaking up Business as Usual." *Built In*, August 9 2021, https://builtin.com/artificial-intelligence/examples-ai-in-industry

8. Marr, Bernard. "What Is Artificial Intelligence (AI) in Business? 10 Practical Examples." *Bernard Marr & Co.*, July 13 2021, https://bernardmarr.com/what-is-artificial-intelligence-ai-in-business-10-practical-examples/

9. Hagan, Shelly. "More Robots Mean 120 Million Workers Need to Be Retrained." *Bloomberg*, Bloomberg, September 5 2019, https://www.bloomberg.com/news/articles/2019-09-06/robots-displacing-jobs-means-120-million-workers-need-retraining

10. Likens, Scott, et al. *AI Predictions 2021*. PwC, February 10 2017, https://www.pwc.com/us/en/tech-effect/ai-analytics/ai-predictions.html

11. Schroer, Alyssa. "50 Artificial Intelligence Companies to Watch in 2022." *Built In*, July 15 2021, https://builtin.com/artificial-intelligence/ai-companies-roundup

Special Insert 1

1. Humanium, "Right to Education: Situation around the world." https://www.humanium.org/en/right-to-education/, *Humanium*.
2. IBID, https://www.humanium.org/en/right-to-education/
3. IBID, https://www.humanium.org/en/right-to-education/
4. IBID, https://www.humanium.org/en/right-to-education/
5. Unesco. "Out-of-School Children and Youth." http://uis.unesco.org/en/topic/out-school-children-and-youth, *UNESCO*.
6. Darling-Hammond, Linda. "Unequal Opportunity: Race and Education." *Brookings*, March 1 1998, https://www.brookings.edu/articles/unequal-opportunity-race-and-education/ .
7. IBID, https://www.brookings.edu/articles/unequal-opportunity-race-and-education/
8. IBID, https://www.brookings.edu/articles/unequal-opportunity-race-and-education/
9. ReFED, "Food Waste." *ReFED*, https://refed.com/food-waste/the-challenge.
10. https://www.un.org/sustainabledevelopment/sustainable-consumption-production/
11. https://www.un.org/en/chronicle/article/losing-25000-hunger-every-day
12. https://www.un.org/en/chronicle/article/losing-25000-hunger-every-day
13. https://www.un.org/en/chronicle/article/losing-25000-hunger-every-day
14. Toossi, Saied. "National School Lunch Program." *U.S. Department of Agriculture*, January 13 2022, https://www.ers.usda.gov/topics/food-nutrition-assistance/child-nutrition-programs/national-school-lunch-program/.
15. FAO, IFAD, UNICEF, WFP, WHO. "The State of Food Security and Nutrition in the World 2020", *Food and Agriculture Organization of the United Nations*, 2020, https://www.fao.org/3/ca9692en/online/ca9692en.html.
16. WHO. "World Bank and WHO: Half the world lacks access to essential health services, 100 million still pushed into extreme poverty because of health expenses." *World Health Organization*, December 13 2017, https://www.who.int/news/item/13-12-2017-world-bank-and-who-half-the-world-lacks-access-to-essential-health-services-100-million-still-pushed-into-extreme-poverty-because-of-health-expenses.
17. IBIDhttps://www.who.int/news/item/13-12-2017-world-bank-and-who-half-the-world-lacks-access-to-essential-health-services-100-million-still-pushed-into-extreme-poverty-because-of-health-expenses
18. IBIDhttps://www.who.int/news/item/13-12-2017-world-bank-and-who-half-the-world-lacks-access-to-essential-health-services-100-million-still-pushed-into-extreme-poverty-because-of-health-expenses
19. Leonard, Jayne. "What is health inequality?" *Medical News Today*, May 16 2021, https://www.medicalnewstoday.com/articles/health-inequity#examples.
20. Ndugga Nambi, Artiga Samantha. "Disparities in Health and Health Care: 5 Key Questions and Answers." *KFF*, May 11 2021, https://www.kff.org/racial-equity-and-health-policy/issue-brief/disparities-in-health-and-health-care-5-key-question-and-answers/.
21. "Health, United States – Infographics." *Centers of Disease Control and Prevention*, April 23 2019, https://www.cdc.gov/nchs/hus/spotlight/2019-heart-disease-disparities.htm.
22. "Health Care Equity: A Call to Action." *Radiological Society of North America*, June 8 2021, https://www.rsna.org/news/2021/june/Health%20Care%20Equity%20Infographic.
23. OXFAM. "World's billionaires have more wealth than 4.6 billion people." *OXFAM International*, January 20 2020, https://www.oxfam.org/en/press-releases/worlds-billionaires-have-more-wealth-46-billion-people.
24. IBID,https://www.oxfam.org/en/press-releases/worlds-billionaires-have-more-wealth-46-billion-people
25. Ricketts Lowell, Kent Ana. "Has Wealth Inequality in America Changed over Time? Here Are Key Statistics." *Federal Reserve Bank of St. Louis*, December 2 2020, https://www.stlouisfed.org/open-vault/2020/december/has-wealth-inequality-changed-over-time-key-statistics.
26. "Income Inequality in the United States." *Inequality.org*, https://inequality.org/facts/income-inequality/.
27. IBID,https://inequality.org/facts/income-inequality/

28. "List of countries by GDP (nominal) per capita." *Wikipedia*, February 15 2022, https://en.wikipedia.org/wiki/List_of_countries_by_GDP_(nominal)_per_capita.
29. Peer, Andrea. "Global poverty: Facts, FAQs, and how to help." *World Vision*, August 23 2021, https://www.worldvision.org/sponsorship-news-stories/global-poverty-facts.
30. "Poverty Rate by Country 2022." *World Population Review*, https://worldpopulationreview.com/country-rankings/poverty-rate-by-country.
31. WHO. "1 in 3 people globally do not have access to safe drinking water – UNICEF,WHO." *World Health Organization*, June 18 2019, https://www.who.int/news/item/18-06-2019-1-in-3-people-globally-do-not-have-access-to-safe-drinking-water-unicef-who.
32. Ritchie Hannah, Roser Max. ""Access to Energy." *Our World in Data*, https://ourworldindata.org/energy-access#access-to-electricity.
33. IBID,https://ourworldindata.org/energy-access#access-to-electricity
34. UNICEF. "Billions of people will lack access to safe water, sanitation and hygiene in 2030 unless progress quadruples—warn WHO, UNICEF." *UNICEF*, July 1 2021, https://www.unicef.org/press-releases/billions-people-will-lack-access-safe-water-sanitation-and-hygiene-2030-unless.
35. IBID,https://www.unicef.org/press-releases/billions-people-will-lack-access-safe-water-sanitation-and-hygiene-2030-unless
36. "The Water Crisis." *Water.org*, https://water.org/our-impact/water-crisis/.
37. IBID,https://water.org/our-impact/water-crisis/
38. "The 17 Goals." *United Nations*, https://sdgs.un.org/goals.

Chapter 4

1. Grannell, Craig. "Best Iphone AR Apps and Games in 2021." *Tom's Guide*, April 14 2021, https://www.tomsguide.com/round-up/best-iphone-ar-apps
2. Zoria, Sophie. *Enterprise AR: 7 Real-World Use Cases for 2021*. AR/VR Journey: Augmented & Virtual Reality Magazine, March 12 2021, https://arvrjourney.com/enterprise-ar-7-real-world-use-cases-for-2021-81ea0319b8e5?gi=f8f1ab935e75
3. *Virtual Reality Market Share & Trends Report, 2021-2028. Grand View Research*, March 2021, https://www.grandviewresearch.com/industry-analysis/virtual-reality-vr-market
4. Davies, Aran. "10 Great Tools for VR Development I DevTeam.space." *DevTeam.Space*, March 16 2021, https://www.devteam.space/blog/10-great-tools-for-vr-development
5. "10 Business Applications of Virtual Reality (VR) Technology." *Sendian Creations*, Sendian Creations, February 8 2021, https://www.sendiancreations.com/top-vr-business-applications
6. "The 9 Best Mixed Reality Headsets of 2021." *Aniwaa*, August 6 2021, https://www.aniwaa.com/buyers-guide/vr-ar/best-mixed-reality-headsets
7. *Virtual Reality Market Share & Trends Report, 2021-2028. Grand View Research*, March 2021, https://www.grandviewresearch.com/industry-analysis/virtual-reality-vr-market
8. *Augmented Reality Market Size & Share Report, 2021-2028. Grand View Research*, February 2021, https://www.grandviewresearch.com/industry-analysis/augmented-reality-market
9. Snider, Mike, and Brett Molina. *USA Today*, November 11 2021, https://www.usatoday.com/story/tech/2021/11/10/metaverse-what-is-it-explained-facebook-microsoft-meta-vr/6337635001.
10. Kafka, Peter. "Facebook Is Quietly Buying up the Metaverse." *Vox*, November 11 2021, https://www.vox.com/recode/22776461/facebook-meta-metaverse-monopoly
11. Warren, Tom. "Microsoft Teams Enters the Metaverse Race with 3D Avatars and Immersive Meetings." *The Verge*, November 2 2021, https://www.theverge.com/2021/11/2/22758974/microsoft-teams-metaverse-mesh-3d-avatars-meetings-features?scrolla=5eb6d68b7fedc32c19ef33b4
12. Thurman, Andrew. "Barbados to Become First Sovereign Nation with an Embassy in the Metaverse." *Yahoo!*, November 15 2021, https://www.yahoo.com/now/barbados-become-first-sovereign-nation-110000022.html
13. Boland, Mike. "How Big Is the Mobile AR Market?" *AR Insider*, March 22 2021, https://arinsider.co/2021/02/23/how-big-is-the-mobile-ar-market-2

Chapter 5

1. "3D Print-Knit." *Ministry of Supply*, https://www.ministryofsupply.com/technologies/3d-print-knit
2. Scott, Clare. "Julia Daviy Uses 3D Printing to Create Beautiful Biodegradable Fashion." *3DPrint. Com*, May 7 2018, https://3dprint.com/212640/julia-daviy-3d-printing
3. *VIP TIE - Sustainable Luxury Accessories - Dubai, Milan*, https://www.viptie3d.com
4. Petch, Michael. "Adidas and Carbon Announce Futurecraft 4D and Plans for Industrial Scale Mass Customization with 3D Printing." *3D Printing Industry*, April 7 2017, https://3dprintingindustry.com/news/adidas-carbon-announce-futurecraft-4d-plans-industrial-scale-mass-customization-3d-printing-110274
5. Plewa, Kat. "3D Printed Jewelry: Why You Should Start Thinking about It?" *Sculpteo*, June 19 2019, https://www.sculpteo.com/blog/2019/06/19/3d-printed-jewelry-why-you-should-start-thinking-about-it
6. Hendrixson, Stephanie. "These 3D Printed Glasses Are Designed to 'Solve' Kids' Eyewear." *Additive Manufacturing*, December 2 2019, https://www.additivemanufacturing.media/articles/these-3d-printed-glasses-are-designed-to-solve-kids-eyewear
7. Fox News. "Chinese Company Builds First 3D-Printed Apartment Building, Mansion." *Fox News*, FOX News Network, January 20 2015, https://www.foxnews.com/tech/chinese-company-builds-first-3d-printed-apartment-building-mansion
8. Rogers, James. "Marines 3D-Print Concrete Barracks in Just 40 Hours." *Fox News*, FOX News Network, August 29 2018, https://www.foxnews.com/tech/marines-3d-print-concrete-barracks-in-just-40-hours
9. Howarth, Dan. "Icon Completes 'First 3D-Printed Homes for Sale in the US.'" *Dezeen*, August 31 2021, https://www.dezeen.com/2021/08/31/east-17th-street-residences-3d-printed-homes-icon-austin
10. Peters, Adele. "This Wild-Looking House Is Made out of Dirt by a Giant 3D Printer." *Fast Company*, April 9 2021, https://www.fastcompany.com/90619146/this-wild-looking-house-is-made-out-of-dirt-by-a-giant-3d-printer
11. "7 Exciting 3D Printed Food Projects Changing How We Eat Forever." *3DSourced*, August 19 2021, https://www.3dsourced.com/guides/3d-printed-food
12. Corolo, Lucas. "3D Printed Food: All You Need to Know in 2021." *All3DP*, November 22 2021, https://all3dp.com/2/3d-printed-food-3d-printing-food
13. Varkey, M., Visscher, D.O., van Zuijlen, P.P.M. *et al.* Skin bioprinting: the future of burn wound reconstruction?. *Burn Trauma 7,* 4 (2019). https://doi.org/10.1186/s41038-019-0142-7
14. *Collplant*, https://collplant.com
15. *Dimension Inx*, https://www.dimensioninx.com
16. *3DBio Therapeutics*, https://3dbiocorp.com
17. *Frontier Bio*, https://www.frontierbio.com
18. *Printivo*, https://www.printivo.eu
19. *Enabling The Future*, https://enablingthefuture.org
20. Gonzalez, Carlos. *The Future of 3D Printing on Human Skin*. Machine Design, May 3 2018, https://www.machinedesign.com/3d-printing-cad/article/21836698/the-future-of-3d-printing-on-human-skin
21. D., Jamie. "Top 12 Best Websites to Download Free STL Files." *3Dnatives*, March 17 2020, https://www.3dnatives.com/en/top-10-websites-stl-files-161120174
22. "Software for 3D Printing." *3D Printing*, https://3dprinting.com/software
23. "Software for 3D Printing." *3D Printing*, https://3dprinting.com/software
24. "3D Printing Market...Global Forecast to 2026." *Markets and Markets* , https://www.marketsandmarkets.com/Market-Reports/3d-printing-market-1276.html
25. https://redetec.com/products/protocycler
26. "3D Printing Market - Growth, Trends, COVID-19 Impact, and Forecasts (2022 - 2027)." *Mordor Intelligence*, https://www.mordorintelligence.com/industry-reports/3d-printing-market

27. "Paper and Paperboard: Material-Specific Data ." *EPA*, Environmental Protection Agency, https://www.epa.gov/facts-and-figures-about-materials-waste-and-recycling/paper-and-paperboard-material-specific-data
28. "3D Printing Market...Global Forecast to 2026." *Markets and Markets* , https://www.marketsandmarkets.com/Market-Reports/3d-printing-market-1276.html

Chapter 6

1. U.S. Department of Transportation. National Highway Traffic Safety Administration. *Critical Reasons for Crashes Investigated in the National Motor Vehicle Crash Causation Survey*, 2015. *DOT HS 812 115*. https://crashstats.nhtsa.dot.gov/Api/Public/ViewPublication/812115
2. U.S. General Services Administration. Office of Motor Vehicle Management. *Crashes Are No Accident*.
3. https://drivethru.gsa.gov/DRIVERSAFETY/DistractedDrivingPosterA.pdf
4. U.S. General Services Administration. Office of Motor Vehicle Management. *Crashes Are No Accident*.
5. https://drivethru.gsa.gov/DRIVERSAFETY/DistractedDrivingPosterA.pdf
6. Lardieri, Alexa. "Traffic Deaths Increased in 2020 Despite Fewer People on Roads During Pandemic." *U.S. News*, June 4 2021, https://www.usnews.com/news/health-news/articles/2021-06-04/traffic-deaths-increased-in-2020-despite-fewer-people-on-roads-during-pandemic
7. U.S. Department of Transportation. National Highway Traffic Safety Administration. *Drunk Driving*.
8. https://www.nhtsa.gov/risky-driving/drunk-driving
9. "SAE Levels of Driving Automation™ Refined for Clarity and International Audience." *SAE International*, May 3 2021, https://www.sae.org/blog/sae-j3016-update
10. "Autonomous Vehicles." *NAIC*, September 22 2021, https://content.naic.org/cipr_topics/topic_autonomous_vehicles.htm
11. Mayer, Tom, et al. *Will This Be the End of Car Dealerships as We Know Them?* KPMG, https://advisory.kpmg.us/content/dam/advisory/en/pdfs/the-end-of-car-dealerships.pdf
12. Research and Markets. "Global Autonomous Cars Market (2020 to 2030) - COVID-19 Growth and Change." *GlobeNewswire*, Research and Markets, May 20 2020, https://www.globenewswire.com/news-release/2020/05/20/2036203/0/en/Global-Autonomous-Cars-Market-2020-to-2030-COVID-19-Growth-and-Change.html
13. "The Automotive LiDAR Market." P 26. *Woodside Capital Partners*, Yole Development, April 2018, https://www.woodsidecap.com/wp-content/uploads/2018/04/Yole_WCP-LiDAR-Report_April-2018-FINAL.pdf
14. Dmitriev, Stan, and Simon Wright. "Autonomous Cars Generate More than 300 TB of Data per Year." *Tuxera*, 2021, https://www.tuxera.com/blog/autonomous-cars-300-tb-of-data-per-year
15. Edmonds, Ellen. "Self-Driving Cars Stuck in Neutral on the Road to Acceptance." *AAA Newsroom*, March 5 2020, https://newsroom.aaa.com/2020/03/self-driving-cars-stuck-in-neutral-on-the-road-to-acceptance

Chapter 7

1. 2022, *Drones Market Size, Share, Growth, Trends And Industry Analysis*, https://brandessenceresearch.com/technology-media/global-drone-market-2018-2024.
2. "Using Drones to Deliver Blood in Rwanda." *BBC NEWS*, March 19 2019, https://www.bbc.com/news/av/business-47631709.
3. "Kroger and Drone Express Partner to Provide Grocery Delivery by Drone." *The Kroger Co.*, March 5 2021, https://ir.kroger.com/CorporateProfile/press-releases/press-release/2021/Kroger-and-Drone-Express-Partner-to-Provide-Grocery-Delivery-by-Drone/default.aspx.
4. Peters, Jay. "UPS Just Won FAA Approval to Fly as Many Delivery Drones as It Wants." *The Verge*, October 1 2019, https://www.theverge.com/2019/10/1/20893655/ups-faa-approval-delivery-drones-airline-amazon-air-uber-eats-alphabet-wing.

5. "Wing Drone Deliveries Take Flight with Fedex®." *FedEx*, October 18 2019, https://www.fedex.com/en-us/sustainability/wing-drones-transport-fedex-deliveries-directly-to-homes.html.
6. "First Prime Air Delivery." *Amazon*, https://www.amazon.com/Amazon-Prime-Air/b?node=8037720011.
7. "Drones for Police, Fire and Public Safety." *Blue Skies Drones*, https://www.blueskiesdronerental.com/first-responders.
8. "Drones for Agriculture." *Precision Hawk*, https://www.precisionhawk.com/agriculture/drones.
9. French, Sally. "What Real Estate Agents Should Know about Drones for Aerial Photography." *The Drone Girl*, October 5 2021, https://www.thedronegirl.com/2021/05/21/real-estate-agents-using-drones.
10. 2022, *Drones Market Size, Share, Growth, Trends And Industry Analysis*, https://brandessenceresearch.com/technology-media/global-drone-market-2018-2024.
11. "CityAirbus NextGen." *Airbus*, https://www.airbus.com/en/innovation/zero-emission/urban-air-mobility/cityairbus-nextgen.
12. Vijayenthiran, Viknesh. "Toyota-Backed Flying Taxi Prototype Takes to the Skies." *Motor Authority*, September 1 2020, https://www.motorauthority.com/news/1129452_toyota-backed-flying-taxi-prototype-takes-to-the-skies.
13. Vijayenthiran, Viknesh. "Toyota-Backed Flying Taxi Startup Joby Aviation Goes Public via SPAC Deal." *Motor Authority*, August 13 2021, https://www.motorauthority.com/news/1131408_toyota-backed-flying-taxi-startup-joby-aviation-goes-public-via-spac-deal.
14. Vijayenthiran, Viknesh. "Hyundai Launches Supernal Flying Taxi Division, Promises First Commercial Flight by 2028." *Motor Authority*, November 10 2021, https://www.motorauthority.com/news/1132611_hyundai-launches-supernal-flying-taxi-division-promises-first-commercial-flight-by-2028.
15. McFarland, Matt. "Boeing's First Autonomous Air Taxi Flight Ends in Fewer than 60 Seconds." *CNN*, January 23 2019, https://www.cnn.com/2019/01/23/tech/boeing-flying-car/index.html.
16. Tuohy, Jennifer Pattison. "Amazon Is Now Accepting Your Applications for Its Home Surveillance Drone." *The Verge*, September 21 2021, https://www.theverge.com/2021/9/28/22692048/ring-always-home-cam-drone-amazon-price-release-date-specs.

Special Insert 2

1. Schwab, Klaus, *The Fourth Industrial Revolution*, Penguin Group, 2013.
2. Schwab, Klaus, *Shaping the Fourth Industrial Revolution*, Currency, 2018.
3. Ries, Eric, *The Lean Startup: How Today's Entrepreneurs Use Continuous Innovation to Create Radically Successful Business*, Currency, 2011.
4. Thiel, Peter and Blake Masters, *Zero to One: Notes on Startups, or How to Build the Future, Currency*, 2014.
5. Christensen, Clayton, *The Innovator's Dilemma: When New Technologies Cause Great Firms to Fail (Management of Innovation and Change)*, Harvard Business Review Press, 2016.
6. Christensen, Clayton, *The Innovator's Solution: Creating and Sustaining Successful Growth*, Harvard Business Review Press, 2013.
7. Christensen, Clayton, *Disrupting Class: How Disruptive Innovation Will Change the Way the World Learns*, McGraw-Hill, 2nd edition, 2016.
8. Christensen, Clayton, *The Innovator's Prescription: A Disruptive Solution for Health Care*, McGraw-Hill, 2016.
9. Bhargava, Rohit, *Non Obvious Megatrends: How to See What Others Miss and Predict the Future*, Ideapress Publishing, 2021.
10. Fired, Jason and David Heinemeier-Hansson, *Rework*, Currency, 2010.
11. Weinberg, Gabriel and Justin Mares, *Traction*, Penguin UK, 2015.
12. Sinek, Simon, *Start with Why: Great Leaders Inspire Everyone to Take Action*, Portfolio, 2011.
13. Hsieh, Tony, *Delivering Happiness*, Grand Central Publishing, 2013.

Chapter 8

1. "U.S. energy facts explained," *U.S. Energy Information Administration*, May 14 2021, https://www.eia.gov/energyexplained/us-energy-facts/.
2. Cox, Chelsey, and Michelle Shen. "What Are the Effects of Climate Change Costing Consumers, on Average?" *USA Today*, Gannett Satellite Information Network, November 2 2021, https://www.usatoday.com/story/money/2021/11/02/what-climate-change-costing-average-american-per-month/8544943002.
3. "Just 1 Second of the Sun's Energy Output Would Power the US for 9,000,000 Years." *Wow Really*, https://www.wowreally.blog/2006/10/ust-1-second-of-suns-energy-output.html#:~:text=Just%201%20second%20of%20the,the%20US%20for%209%2C000%2C000%20years.
4. Bekele, Adisu, et al. "Large-Scale Solar Water Heating Systems Analysis in Ethiopia: A Case Study." *International Journal of Sustainable Energy*, vol. 32, no. 4, September 23 2011, pp. 207–228., https://doi.org/10.1080/14786451 .2011.605951.
5. Chandler, David L. "Simple, Solar-Powered Water Desalination." *MIT News*, February 6 2020, https://news.mit.edu/2020/passive-solar-powered-water-desalination-0207.
6. *Household Water Treatment Options in Developing Countries: Solar Disinfection (SODIS)*. Centers for Disease Control and Prevention & US Aid, January 2008, https://web.archive.org/web/20080529090729/http://www.ehproject.org/PDF/ehkm/cdc-options_sodis.pdf.
7. "Hydrogen Explained - Use of Hydrogen." U.S. Energy Information Administration (EIA), January 20 2022, https://www.eia.gov/energyexplained/hydrogen/use-of-hydrogen.php.
8. D'Allegro, Joe. "Elon Musk Says the Tech Is 'Mind-Bogglingly Stupid,' but Hydrogen Cars May Yet Threaten Tesla." *CNBC*, February 24 2019, https://www.cnbc.com/2019/02/21/musk-calls-hydrogen-fuel-cells-stupid-but-tech-may-threaten-tesla.html.
9. Schwartz, Ariel. "A French Sidewalk Lets You Power the Streetlights With Your Feet." *Fast Company*, April 15 2010, https://www.fastcompany.com/1617178/french-sidewalk-lets-you-power-streetlights-your-feet.
10. *Solar Marketplace Intel Report* (Report) (13th ed.), *EnergySage,* August 2021.
11. Cohen, Ariel. "What Batteries Will Power the Future?" *Forbes*, Forbes Magazine, February 11 2021, https://www.forbes.com/sites/arielcohen/2021/02/11/what-batteries-will-power-the-future/?sh=53bb21bb41c0.

Chapter 9

1. "Automatic Speech Recognition (ASR) Systems Compared." *Medium*, Sciforce, July 7 2021, https://medium.com/sciforce/automatic-speech-recognition-asr-systems-compared-6ad5e54fd65f.
2. "GnoSys App Translates Sign Language into Speech in Real Time Using the Power of AI." *Newz Hook*, October 28 2018, https://newzhook.com/story/20387.

Chapter 10

1. "What Are the Differences between 2G, 3G, 4G LTE, and 5G Networks?" *RantCell*, https://rantcell.com/comparison-of-2g-3g-4g-5g.html.

Chapter 11

1. Metz, Cade. "Google Claims a Quantum Breakthrough That Could Change Computing." *The New York Times*, October 23 2019, https://www.nytimes.com/2019/10/23/technology/quantum-computing-google.html.
2. Outeiral, Carlos, et al. WIREs, 2020, *The Prospects of Quantum Computing in Computational Molecular Biology*, https://doi.org/10.1002/wcms.1481.2.
3. Tangermann, Victor. "This Quantum Desktop Computer Can Be Yours for Just $5,000." *Futurism*, February 5 2021, https://futurism.com/the-byte/quantum-desktop-computer-5000.

4. Bradley, Michael. "Google Claims to Have Invented a Quantum Computer, but IBM Begs to Differ." *The Conversation*, August 10 2020, https://theconversation.com/google-claims-to-have-invented-a-quantum-computer-but-ibm-begs-to-differ-127309.
5. "Quantum Computing Is Now Available on AWS through Amazon Braket." *Amazon*, 2018, https://aws.amazon.com/about-aws/whats-new/2020/08/quantum-computing-available-aws-through-amazon-braket.
6. Ranger, Steve. "What Is Cloud Computing? Everything You Need to Know about the Cloud Explained." *ZDNet*, December 13 2018, https://www.zdnet.com/article/what-is-cloud-computing-everything-you-need-to-know-about-the-cloud.
7. Chand, Mahesh. "Top 10 Cloud Service Providers in 2021." *C# Corner*, https://www.c-sharpcorner.com/article/top-10-cloud-service-providers.
8. "Nanotechnology and Nanoparticles in Drug Delivery." *UnderstandingNano*, https://www.understandingnano.com/nanotechnology-drug-delivery.html.
9. Freitas, Robert A. "The Ideal Gene Delivery Vector: Chromallocytes, Cell Repair Nanorobots for Chromosome Replacement Therapy." *Journal of Evolution and Technology*, vol. 16, no. 1, June 2007, pp. 1–97. *ISSN 1541-0099*.

The Takedown

1. Moughal, Jamil. "Top Tech Skills for the Future of Work." *Nerd for Tech*, May 23 2021, https://medium.com/nerd-for-tech/top-tech-skills-for-the-future-of-work-abcec23a1a3d.
2. "18 In-Demand Technology Skills to Learn in 2022." *Learn to Code with Me*, January 3 2022, https://learntocodewith.me/posts/tech-skills-in-demand/.
3. Chorna, Iryna. "A Guide to Future-Oriented Skills: Skills in Demand to Watch in the Next Five Years," *HRForecast*, October 7 2021, https://hrforecast.com/a-guide-to-future-oriented-skills-skills-in-demand-to-watch-in-the-next-five-years/.
4. "51 In-Demand Tech Skills for Technology Careers," *indeed*, September 23 2021, https://www.indeed.com/career-advice/career-development/in-demand-tech-skills.
5. Columbus, Louis. "Top 10 Tech Job Skills Predicted to Grow the Fastest in 2021," December 27 2020, *Forbes*, https://www.forbes.com/sites/louiscolumbus/2021/12/27/top-10-tech-job-skills-predicted-to-grow-the-fastest-in-2021/.
6. Stevens-Huffman, Leslie. "Top 10 IT Skills In-Demand for 2021," March 8 2022, *ITCareerFinder*, https://www.itcareerfinder.com/brain-food/blog/entry/top-10-technology-skills-2021.html.

Index – Definitions

About the Author
STEPHEN HAAG

Assuming you've read the book, you already know a lot about me. Here are the more formal details.

Stephen Haag is a Professor of the Practice in Daniels College of Business Department of Business Information and Analytics at the University of Denver. Stephen received his MBA from West Texas State University in 1988 and his PhD from the University of Texas at Arlington in 1992. After joining the University of Denver in 1995, Stephen has held the positions of Chair of the Department of Information Technology and Electronic Commerce, the Director of Daniels Technology, the Director of the MBA, Associate Dean of Graduate Programs, the Director of Assurance of Learning, and the Director of Entrepreneurship.

Stephen is the author or co-author of 48 books.

He lives in Highlands Ranch, Colorado with his wife Pam, four children (Darian, Trevor, Katrina, and Alexis), and 2 rescue shelter dogs (Loki and Cricket).

Printed in the USA
CPSIA information can be obtained
at www.ICGtesting.com
LVHW020023260823
756279LV00002B/33